Modern Crop Protection

Modern Crop Protection

Remi Ford

MURPHY & MOORE
www.murphy-moorepublishing.com

Murphy & Moore Publishing,
1 Rockefeller Plaza,
New York City, NY 10020, USA

Visit us on the World Wide Web at:
www.murphy-moorepublishing.com

ISBN: 978-1-63987-371-5 (Hardback)

Cataloging-in-Publication Data

Modern crop protection / Remi Ford.
p. cm.
Includes bibliographical references and index.
ISBN 978-1-63987-371-5
1. Plants, Protection of. 2. Agriculture. 3. Plant protection industry.
I. Ford, Remi.

SB950 .M63 2022
632.9--dc23

Table of Contents

Preface

The branch of science which studies and analyses the methods involved in protecting the crops from different agents such as pests, weeds, and other organisms which cause damage to them is called crop protection. Some of the techniques of crop protection are pests and insects management, weed management and plant disease management. Weed management involves techniques such as spraying of herbicides, weedicides, ploughing of field, etc. Pest management involves spraying of pesticides in the fields to minimize crop damage. Plant disease management refers to controlling the growth of microorganisms like fungi, viruses and bacteria which affect the different parts of a crop. This is done with the help of biocontrol agents like wasps, beetles and dragonflies, and using techniques like threshing and winnowing. Crop protection practices vary from crop to crop. The biological characteristics of a crop are key factors in deciding the suitable management technique. The topics included in this book on crop protection are of utmost significance and bound to provide incredible insights to readers. It is a valuable compilation of topics, ranging from the basic to the most complex theories and principles in the field of crop protection. This textbook is appropriate for those seeking detailed information in this area.

A detailed account of the significant topics covered in this book is provided below:

Chapter 1- Any plant or animal product which is grown and harvested on a large scale commercially for profit is termed as a crop. There are various types of cropping systems such as monocropping, intercropping, multicropping. This is an introductory chapter which will introduce briefly all the significant aspects of crops and crop protection.

Chapter 2- Crop diseases refer to any alteration or harmful deviation in the normal functioning of the physiological processes of crops. This chapter discusses in detail the diseases in various crops such as diseases in lentils, diseases in chickpeas, diseases in field peas and diseases in wheat, etc.

Chapter 3- The substances which are used to kill insects are known as insecticides. They can be grouped into three categories: inorganic insecticides, organochloride insecticides and organophosphate insecticides. All these types of insecticides are thoroughly discussed in this chapter.

Chapter 4- The substances that stop the growth of pests are termed as pesticides. Some of the various types of pesticides are herbicide, molluscicide, acaricide, piscicide and dicofol. Herbicides are the substances which are used to control the growth of unwanted plants. The chapter closely examines these types of pesticides to provide an extensive understanding of the subject.

Chapter 5- There are various methods and techniques of weed and pest control. Some of these are mechanical weed control, bush regeneration, soil steam sterilization, soil solarization, etc. This chapter discusses in detail these methods and techniques, related to weed and pest control.

It gives me an immense pleasure to thank our entire team for their efforts. Finally in the end, I would like to thank my family and colleagues who have been a great source of inspiration and support.

Remi Ford

Chapter 1

Introduction to Crop Protection

Any plant or animal product which is grown and harvested on a large scale commercially for profit is termed as a crop. There are various types of cropping systems such as monocropping, intercropping, multicropping. This is an introductory chapter which will introduce briefly all the significant aspects of crops and crop protection.

Crops

In agriculture, Crop is a plant or plant product that can be grown and harvested extensively for profit or subsistence. By use, crops fall into six categories: food crops, for human consumption (e.g., wheat, potatoes); feed crops, for livestock consumption (e.g., oats, alfalfa); fibre crops, for cordage and textiles (e.g., cotton, hemp); oil crops, for consumption or industrial uses (e.g., cottonseed, corn); ornamental crops, for landscape gardening (e.g., dogwood, azalea); and industrial and secondary crops, for various personal and industrial uses (e.g., rubber, tobacco).

Classification of Crops

Field crops are herbaceous plants grown in cultivated fields under more or less extensive system of culture. Agronomic classification of different field crop is given as under:

According to Botanical Aspects

- Monocotyledons: A plant having only one seed leaf or cotyledon in each of its seed. Such plants have a shallow roots, narrow leaves and tillering ability. e.g., Families:
    - Poaceae, e.g. Cereals i.e. paddy, wheat, sorghum, pearl millet, maize, sugarcane, etc.
    - Zingiberaceae, e.g. Ginger.
    - Liliaceae, e.g. Onion.
- Dicotyledons: A plant having two seed leaf or cotyledons in each of its seed.

Such plants have a deep taproots, broad leaves and branching ability. e.g., Families:

- Malvaceae, e.g. cotton, okra.
- Cruciferae, e.g. cabbage, mustard.
- Solanaceae, e.g. tobacco, potato, brinjal.
- Leguminoceae, e.g. groundnut, pigeon pea, gram, pea, mung bean, urd bean.
- Tiliaceae, e.g. jute, sun hemp.
- Linaceae, eg. linseed.
- Euphorbiaceae, e.g. castor.
- Composite, e.g. sunflower.
- Chenopodiaceae, e.g. sugar beet.

According to Season of Growth

- Kharif or Monsoon: Crops are grown in the month of June - July. e.g. Paddy, pigeon pea, groundnut, sorghum, etc.
- Rabi or Winter: Crops are grown in the month of October - November. e.g. wheat, gram, mustard, cumin, fenugreek, onion etc.
- Summer or Hot Weather: Crops are grown in the month of February - June. e.g., paddy, green gram, black gram, cowpea, etc.

According to Life Period

- Annuals: A plant which grows from seed, complete its life cycle and dies during the season or same year. E.g. cereal crops.
- Biennials: A plant which completes its life cycle in two season or year. During first season/year, they complete its vegetative growth and during second season/year they completes its reproductive growth. eg. Sugarcane, sugar beet, banana, onion.
- Perennial: A plant which completes its life cycle in more than two seasons/ years. e.g., agave, elephant foot.

According to Cultural Practices

- Irrigated crops, e.g. Sugarcane, paddy, banana etc.

- Dry farming crops: The crop which are grown under natural precipitation. E.g. sorghum, bajra, groundnut, nagli, etc.
- Sole crops: One crop variety grown alone in pure stands at normal density.
- Monocroping: The repetitive growing of the same crops on the same piece of land. E.g. groundnut in Saurastra region.
- Intercropping: Growing two or more crops simultaneously on the same piece of land in different rows e.g., sorghum + black gram, pigeon pea + ground nut, sugarcane + onion, etc.
- Mixed cropping: Growing two or more crops simultaneously on the same land without/ irrespective to definite row pattern. E.g. sorghum + green gram+ black gram.

According to Agronomical or Economical Aspects

Food crops:

- Cereals, e.g. paddy, wheat, maize, sorghum, bajra, nagli, etc.
- Pulses, e.g. gram, green gram, black gram, pigeon pea, cowpea, moth bean, etc.
- Legumes, e.g. groundnut, soybean, cluster bean.
- Edible oil seeds, e.g. groundnut, seasamum, sunflower, mustard, safflower.
- Fruit crops, e.g. mango, banana, sapota, guava, papaya, ber, grape citrus crop, etc.
- Vegetable crops:
    - Leafy: Cabbage, fenugreek, palak.
    - Fruit: Tomato, okra, brinjal.
    - Root: Radish, carrot.
    - Tuber/Stem: potato, sweet potato, ginger, turmeric.
    - Bulb: Onion, garlic.
    - Flower: Cauliflower.
    - Pod: French bean, pigeon pea, cowpea.

Non-food crops:

- Forage/fodder: lucerne, berseem, hybrid nappier grass.

- Fibre crop: Cotton, jute, sun hemp.
- Non edible oilseeds: Castor, linseed.
- Sugar crop: Sugarcane, sugar beet.
- Dyes: Safflower, indigo.
- Narcotics: Tobacco, coffee, tea, opium, poppy, chicory.
- Drugs/Medicinal: Isabgul, jethimadh, senna, kariyatu.
- Spices & condiments: Cardamom, cumin, black pepper, coriander, fennel, fenugreek.

Special purpose crops:

- Row crops: Crops which are grown in rows with uniform spacing throughout the field. e. g. cotton, castor, sorghum, etc.
- Support crops: Certain fast growing crops work as supporter to vine crops. e.g. castor, shevri in betel vine, sorghum in cowpea/bean.
- Wind break crops: Crops which are grown on boundaries to protect the field crops from wind.
- Cover crops: Cover crops are grown primarily to cover the soil and to reduce the cost of moisture and erosion by wind and water. E.g. groundnut, kidney bean, cowpea, mung bean.
- Silage crops: Crop which are preserved in a succulent condition by partial fermentation in a tight silo pit. E.g. maize, sorghum, bajra.
- Cash crops: Crop grown for sale and brings money immediately. E.g. cotton, tobacco, potato, sugarcane.
- Green manure Crops: Any crop which are grown and buried into the soil for improving the soil condition by the addition of organic matter. E.g. sun hemp, dhaincha, glyricidia.
- Pasture crops: Different types of vegetation found on pastures or grassland area which usually grow. e.g. dharo, zinzvo.
- Catch crops: Crop which is grown as substitute for the main crop that has failed on account of unfavorable condition. e.g. cowpea, sesame, green gram.
- Trap crops: Crop which is grown on boundary of the field for protection against pest, insect, disease.

- Nurse crops: Crop which is used to protect or nurse the other crops in their young stage. e.g. cluster bean in ginger, sun hemp in sugarcane.
- Companion crops: Two crops are taken together with the aim that they are benefited to each other. e.g. maize and green gram.
- Mixed crops: Two or more than two crops are grown together on the same piece of land at the same time. e.g. bajra + cowpea + green gram.

Crop Protection

Crop protection is the general method or the practice of protecting the crop yields from different agents including pests, weeds, plant diseases, and other organisms that cause damage to the agricultural crops.

Apart from crops, agricultural fields would have weeds, small animals like rats, mites, insects, pests, disease-causing pathogens and frequently raided by birds. All these factors are mainly responsible for the loss or damage to the crops. Thus to yield high crop production, farmers need to protect the crop from these pests. Hence crop protection management is important before, during and after the cultivation.

Methods of Crop Protection

There are many crop protection tools and practices, which farmers can implement to increase the success of their crops.

Crop Protection before and During the Cultivation

Weed Management

Weeds are unwanted plants growing along with the crops. These undesirable plants, steal the nutrients, sunlight, water and other resources from the crops and affect their growth, which results in the undernourished of crops and decreases the yields. To safeguard the productivity of crops, farmers remove these weeds by a process called weeding. Weeding is the process of controlling the growth of weeds. There are various methods of weeding:

- Spraying weedicides on the weeds.
- Manually plucking the weeds by hands.
- Removing weeds by trowel and harrow.
- Ploughing the field to remove the weeds even before sowing the seeds.

Few examples of weeds are Amaranthus, Cyperinus rotundus, Bermuda grass, etc. Apart from weeding, Herbicides – a chemical substance also play an important role in controlling the growth of the weeds and also help in preventing soil erosion and water loss.

Pests and Insects Management

Both insects and pests are the major cause of crop damage and yield loss. They could ruin the whole crop and eat up the large portion of grains. In fact, they can reduce crop output by 30-50(%) every year if left unchecked. The best ways to protect crop damage are by incorporating integrated pest and insect management. Spraying insecticides, pesticides help to minimize the crop damage by controlling the insects and other pests.

Plant Disease Management

Pathogenic diseases are another type of crop-damaging factors. Microorganisms like bacteria, fungi, and viruses affect different parts of the crop through various means. Crops could be protected from these diseases by using pesticides and biocontrol agents. These are crop protection management, which should be followed before and during the cultivation.

Crop Protection after the Cultivation

Once the crop has reached maturity, it now needs to be harvested. So harvesting is the cutting and gathering of the matured crop. The term harvesting also generally includes the immediate post-harvest practices such as threshing and winnowing. Let us look at both these processes:

- Threshing: The process of separating the grains from their chaffs or pods is threshing.

- Winnowing: After threshing, we must separate the grains from the chaffs. Winnowing is the process of separating the grains.

There are various factors to consider before the harvesting of crops. The crops need close examination to ensure that harvesting is not premature. This leads to shedding of seeds and loss of crop. If the crops are over-ripened, they may lose their value in the market or may even be unconsumable.

Harvesting is generally done manually. Sicles is a tool that is used to cut the crop. This method is a laborious and time-consuming and only suitable for small-sized farms. On larger farms, a harvester is used which combines harvesting with threshing and winnowing.

Cultivated grains must be stored safely to avoid product loss. Chances of grain loss are much higher during this time than before cultivation. Therefore, methods of protection need to be better. Common pests and rodents and some environmental conditions like humidity and temperature are the responsible factors for the loss. Certain precautions help us to prevent this loss. Before storing the grains, it should be cleaned and dried completely in sunlight as it helps in protecting the crops from fungal growth, which is caused by the presence of moisture. Pests could also be killed by a method called fumigation. According to modern agricultural technologies, there are various tools for managing weeds, insects, and disease. The tools include the applications of pesticides, soil profile, nutrient management etc.

Pesticides

Pesticides are chemical substances that are meant to kill pests. In general, a pesticide is a chemical or a biological agent such as a virus, bacterium, antimicrobial, or disinfectant that deters, incapacitates, kills, pests.

This use of pesticides is so common that the term pesticide is often treated as synonymous with plant protection product. It is commonly used to eliminate or control a variety of agricultural pests that can damage crops and livestock and reduce farm productivity. The most commonly applied pesticides are insecticides to kill insects, herbicides to kill weeds, rodenticides to kill rodents, and fungicides to control fungi, mould, and mildew.

Pesticides are not recent inventions. Many ancient civilizations used pesticides to protect their crops from insects and pests. Ancient Sumerians used elemental sulfur to protect their crops from insects. Whereas, Medieval farmers experimented with chemicals using arsenic, lead on common crops.

The Chinese used arsenic and mercury compounds to control body lice and other pests. While, the Greeks and Romans used oil, ash, sulfur, and other materials to protect themselves, their livestock, and their crops from various pests.

Meanwhile, in the nineteenth century, researchers focused more on natural techniques involving compounds made with the roots of tropical vegetables and chrysanthemums. In 1939, Dichloro-Diphenyl-Trichloroethane (DDT) was discovered, which has become extremely effective and rapidly used as the insecticide in the world. However, twenty years later, due to biological effects and human safety, DDT has been banned in almost 86 countries.

The Food and Agriculture Organization (FAO) has defined pesticide as: any substance or mixture of substances intended for preventing, destroying or controlling any pest, including vectors of human or animal disease, unwanted species of plants or animals, causing harm during or otherwise interfering with the production, processing, storage, transport, or marketing of food, agricultural commodities, wood and wood products or animal feedstuffs, or substances that may be administered to animals for the control of insects, arachnids, or other pests in or on their bodies.

Types of Pesticides

These are grouped according to the types of pests which they kill.

- Insecticides – Insects
- Herbicides – Plants
- Rodenticides – Rodents (Rats & Mice)
- Bactericides – Bacteria
- Fungicides – Fungi
- Larvicides – Larvae

Pesticides can also be considered as:

- Biodegradable: The biodegradable kind is those which can be broken down by microbes and other living beings into harmless compounds.
- Persistent: While the persistent ones are those which may take months or years to break down.

Another way to classify these is to consider those that are chemical forms or are derived from a common source or production method.

Chemically-Related Pesticides

- Organophosphate: Most organophosphates are insecticides, they affect the nervous system by disrupting the enzyme that regulates a neurotransmitter.
- Carbamate: Similar to the organophosphorus pesticides, the carbamate pesticides also affect the nervous system by disrupting an enzyme that regulates the neurotransmitter. However, the enzyme effects are usually reversible.
- Organochlorine insecticides: They were commonly used earlier, but now many countries have been removed Organochlorine insecticides from their market due to their health and environmental effects and their persistence (e.g., DDT, chlordane, and toxaphene).
- Pyrethroid: These are a synthetic version of pyrethrin, a naturally occurring pesticide, found in chrysanthemums(Flower). They were developed in such a way as to maximise their stability in the environment.
- Sulfonylurea herbicides: The sulfonylureas herbicides have been commercialized for weed control such as pyrithiobac-sodium, cyclosulfamuron, bispyribac-sodium, terbacil, sulfometuron-methyl Sulfosulfuron, rimsulfuron, pyrazosulfuron-ethyl, imazosulfuron, nicosulfuron, oxasulfuron, nicosulfuron, flazasulfuron, primisulfuron-methyl, halosulfuron-methyl, flupyrsulfuron-methyl-sodium, ethoxysulfuron, chlorimuron-ethyl, bensulfuron-methyl, azimsulfuron, and amidosulfuron.

- Biopesticides: The biopesticides are certain types of pesticides derived from such natural materials as animals, plants, bacteria, and certain minerals.

Examples of Pesticides

Examples of pesticides are fungicides, herbicides, and insecticides. Examples of specific synthetic chemical pesticides are glyphosate, Acephate, Deet, Propoxur, Metaldehyde, Boric Acid, Diazinon, Dursban, DDT, Malathion, etc.

Benefits of Pesticides

The major advantage of pesticides is that they can save farmers. By protecting crops from insects and other pests. However, below are some other primary benefits of it:

- Controlling pests and plant disease vectors.
- Controlling human/livestock disease vectors and nuisance organisms.
- Controlling organisms that harm other human activities and structures.

Effects of Pesticides

The toxic chemicals in these are designed to deliberately release into the environment. Though each pesticide is meant to kill a certain pest, a very large percentage of pesticides reach a destination other than their target. Instead, they enter the air, water, sediments, and even end up in our food.

Pesticides have been linked with human health hazards, from short-term impacts such as headaches and nausea to chronic impacts like cancer, reproductive harm. The use of these also decreases the general biodiversity in the soil. If there are no chemicals in the soil there is higher soil quality, and this allows for higher water retention, which is necessary for plants to grow.

Insecticides

Insecticides are agents of chemical or biological origin that control insects. Control may result from killing the insect or otherwise preventing it from engaging in behaviors deemed destructive. Insecticides may be natural or manmade and are applied to target pests in a myriad of formulations and delivery systems (sprays, baits, slow-release diffusion, etc.). The science of biotechnology has, in recent years, even incorporated bacterial genes coding for insecticidal proteins into various crop plants that deal death to unsuspecting pests that feed on them.

Some 10,000 species of the more than 1 million species of insects are crop-eating, and of these, approximately 700 species worldwide cause most of the insect damage to man's crops, in the field and in storage.

Humanoids have been on earth for more than 3 million years, while insects have existed for at least 250 million years. We can guess that among the first approaches used by our primitive ancestors to reduce insect annoyance was hugging smoky fires or spreading mud and dust over their skin to repel biting and tickling insects, a practice resembling the habits of elephants, swine, and water buffalo. Today, such approaches would be classed as repellents, a category of insecticides.

Historians have traced the use of pesticides to the time of Homer around 1000 B.C., but the earliest records of insecticides pertain to the burning of "brimstone" (sulfur) as a fumigant. Pliny the Elder recorded most of the earlier insecticide uses in his Natural History. Included among these were the use of gall from a green lizard to protect apples from worms and rot. Later, we find a variety of materials used with questionable results: extracts of pepper and tobacco, soapy water, whitewash, vinegar, turpentine, fish oil, brine, lye among many others.

At the beginning of World War II, our insecticide selection was limited to several arsenicals, petroleum oils, nicotine, pyrethrum, rotenone, sulfur, hydrogen cyanide gas, and cryolite. It was World War II that opened the Modern Era of Chemical control with the introduction of a new concept of insect control --synthetic organic insecticides, the first of which was DDT.

Organochlorines

The organochlorines are insecticides that contain carbon (thus organo-), hydrogen, and chlorine. They are also known by other names: chlorinated hydrocarbons, chlorinated organics, chlorinated insecticides, and chlorinated synthetics. The organochlorines are now primarily of historic interest, since few survive in today's arsenal.

Diphenyl Aliphatics

The oldest group of the organochlorines is the diphenyl aliphatics, which included

DDT, DDD, dicofol, ethylan, chlorobenzilate, and methoxychlor. DDT is probably the best known and most notorious chemical of the 20th century. It is also fascinating, and remains to be acknowledged as the most useful insecticide developed. More than 4 billion pounds of DDT were used throughout the world, beginning in 1940, and in the U.S. ending essentially in 1973, when the U.S. Environmental Protection Agency canceled all uses. The remaining First World countries rapidly followed suit. DDT is still effectively used for malaria control in several third world countries. In 1948, Dr. Paul Muller, a Swiss entomologist, was awarded the Nobel Prize in Medicine for his lifesaving discovery of DDT as an insecticide useful in the control of malaria, yellow fever and many other insect-vectored diseases.

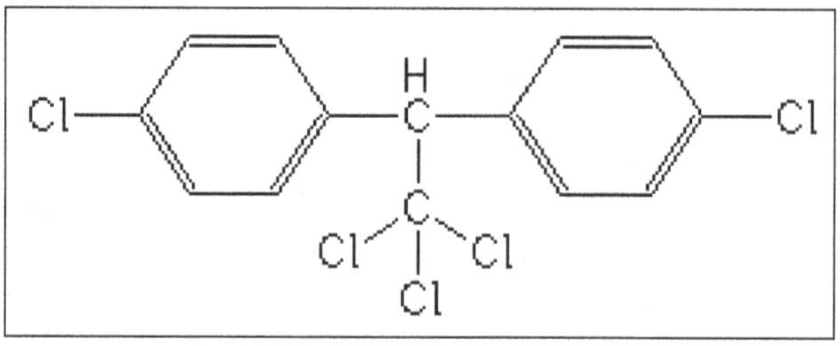

Mode of Action

The mode of action for DDT has never been clearly established, but in some complex manner it destroys the delicate balance of sodium and potassium ions within the axons of the neuron in a way that prevents normal transmission of nerve impulses, both in insects and mammals. It apparently acts on the sodium channel to cause "leakage" of sodium ions. Eventually the affected neurons fire impulses spontaneously, causing the muscles to twitch- "DDT jitters"- followed by convulsions and death. DDT has a negative temperature correlation--the lower the surrounding temperature the more toxic it becomes to insects.

Hexchlorocyclohexane (HCH)

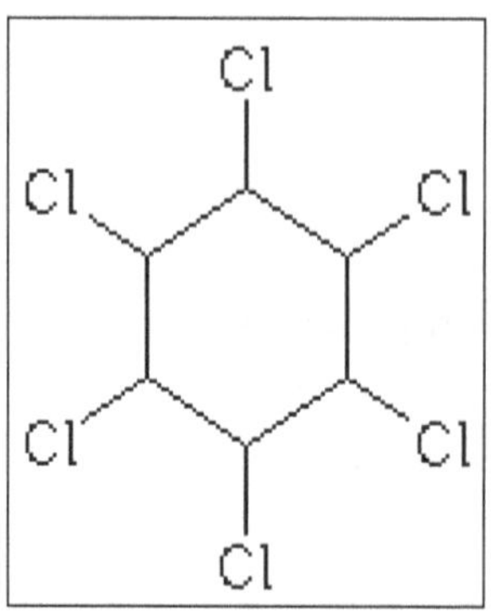

Also known as benzenehexachloride (BHC), the insecticidal properties of HCH were discovered in 1940 by French and British entomologists. In its technical grade, there are five isomers, alpha, beta, gamma, delta and epsilon. Surprisingly, only the gamma isomer has insecticidal properties. Consequently, the gamma isomer was isolated in manufacture and sold as the odorless insecticide lindane. In contrast, technical grade HCH has a strong musty odor and flavor, which can be imparted to treated crops and animal products. Because of its very low cost, HCH is still used in many developing countries. In 2002, the U.S. EPA removed all food-related (tolerance-requiring) uses of lindane from the U.S.

Mode of Action

The effects of HCH superficially resemble those of DDT, but occur much more rapidly, and result in a much higher rate of respiration in insects. The gamma isomer is a neurotoxicant whose effects are normally seen within hours as increased activity, tremors, and convulsions leading to prostration. It too, exhibits a negative temperature correlation, but not as pronounced as that of DDT.

Cyclodienes

The cyclodienes appeared after World War II: chlordane, 1945, aldrin and dieldrin, 1948; heptachlor, 1949; endrin, 1951; mirex, 1954; endosulfan, 1956; and chlordecone, 1958. There were other cyclodienes of minor importance developed in the U.S. and Germany. Most of the cyclodienes are persistent insecticides and are stable in soil and relatively stable to the ultraviolet of sunlight. As a result, they were used in greatest quantity as soil insecticides (especially chlordane, heptachlor, aldrin, and dieldrin) for the control of termites and soil-borne insects whose larval stages feed on the roots of plants.

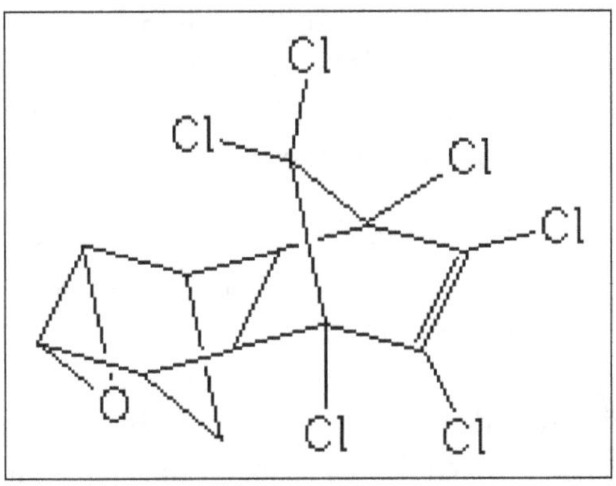

To appreciate the effectiveness of these materials as termiticides, consider that wood and wooden structures treated with chlordane, aldrin, and dieldrin in the year of their development are still protected from damage—after more than 60 years! The cyclodienes were the most effective, long-lasting and economical termiticides ever developed. Because of their persistence in the environment, resistance that developed in several soil insect pests, and in some instances biomagnification in wildlife food chains, most agricultural uses of cyclodienes were canceled by the EPA between 1975 and 1980, and their use as termiticides canceled in 1984-88.

Mode of Action

Unlike DDT and HCH, the cyclodienes have a positive temperature correlation--their toxicity increases with increasing ambient temperature. Their modes of action are also not clearly understood. However, it is known that this group acts on the inhibitory mechanism called the GABA (g-aminobutyric acid) receptor. This receptor operates by increasing chloride ion permeability of neurons. Cyclodienes prevent chloride ions from entering the neurons, and thereby antagonize the "calming" effects of GABA. Cyclodienes appear to affect all animals similarly, first with the nervous activity followed by tremors, convulsions and prostration.

Polychloroterpenes

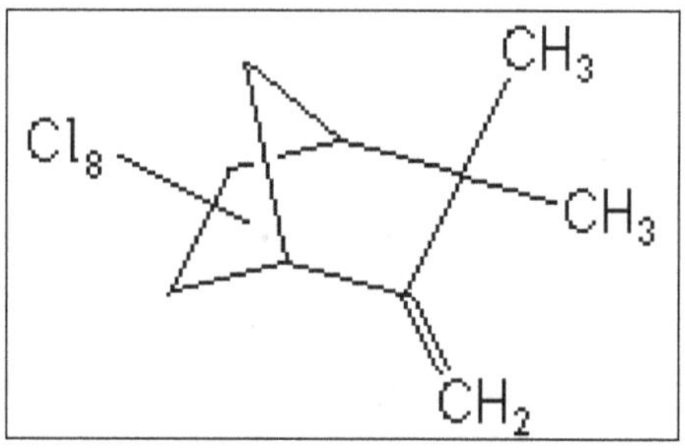

Only two polychloroterpenes were developed toxaphene in 1947, and strobane in 1951. Toxaphene had by far the greatest use of any single insecticide in agriculture, while strobane was relatively insignificant. Toxaphene was used on cotton, first in combination with DDT, for alone it had minimal insecticidal qualities. Then, in 1965, after several major cotton insects became resistant to DDT, toxaphene was formulated with methyl parathion, an organophosphate insecticide.

Toxaphene is a mixture of more than 177 10-carbon polychlorinated derivatives. These materials persist in the soil, though not as long as the cyclodienes, and disappear from the surfaces of plants in 3-4 weeks. This disappearance is attributed more to volatility than to photolysis or plant metabolism. Toxaphene is rather easily metabolized by mammals and birds, and is not stored in body fat nearly to the extent of DDT, HCH and the cyclodienes. Despite its low toxicity to insects, mammals and birds, fish are highly susceptible to toxaphene poisoning, in the same order of magnitude as to the cyclodienes. Toxaphene's registrations were canceled by EPA in 1983.

Mode of Action

Toxaphene and strobane act on the neurons, causing an imbalance in sodium and potassium ions, similar to that of the cyclodiene insecticides.

Organophosphates

Organophosphates (OPs) are the term that includes all insecticides containing phosphorus. Other names used, but no longer in vogue, are organic phosphates, phosphorus insecticides, nerve gas relatives, and phosphoric acid esters. All organophosphates are derived from one of the phosphorus acids, and as a class are generally the most toxic of all pesticides to vertebrates. Because of the similarity of OP chemical structures to the "nerve gases," their modes of action are also similar. Their insecticidal qualities were first observed in Germany during World War II in the study of the extremely toxic OP nerve gases sarin, soman, and tabun. Initially, the discovery was made in search of substitutes for nicotine, which was heavily used as an insecticide but in short supply in Germany.

The OPs have two distinctive features: they are generally more toxic to vertebrates than other classes of insecticides, and most are chemically unstable or non-persistent. It is this latter characteristic that brought them into agricultural use as substitutes for the persistent organochorines. Because of the relatively high toxicity of the OP's, EPA, under provisions of the Food Quality Protection Act, undertook an extensive reappraisal of the entire class beginning in the late 1990's. Many OP's were voluntarily canceled and others lost uses.

Mode of Action

The OPs work by inhibiting certain important enzymes of the nervous system, namely cholinesterase (ChE). The enzyme is said to be phosphorylated when it becomes attached to the phosphorous moiety of the insecticide, a binding that is irreversible. This inhibition results in the accumulation of acetylcholine (ACh) at the neuron/neuron and neuron/muscle (neuromuscular) junctions or synapses, causing rapid twitching of voluntary muscles and finally paralysis.

Classification

All OPs are estersof phosphorus having varying combinations of oxygen, carbon, sulfur and nitrogen attached, resulting in six different subclasses: phosphates, phosphonates, phosphorothioates, phosphorodithioates, phosphorothiolates and phosphoramidates. These subclasses are easily identified by their chemical names. The OPs are generally divided into three groups--aliphatic, phenyl, and heterocyclic derivatives.

Aliphatics

The aliphatic OPs are carbon chain-like in structure. The first OP brought to agriculture, TEPP belonged to this group. Other examples are malathion, trichlorfon (Dylox), monocrotophos (Azodrin), dimethoate (Cygon), oxydemetonmethyl (Meta Systox), dicrotophos (Bidrin), disulfoton (Di-Syston), dichlorvos (Vapona), mevinphos (Phosdrin), methamidophos (Monitor), and acephate (Orthene).

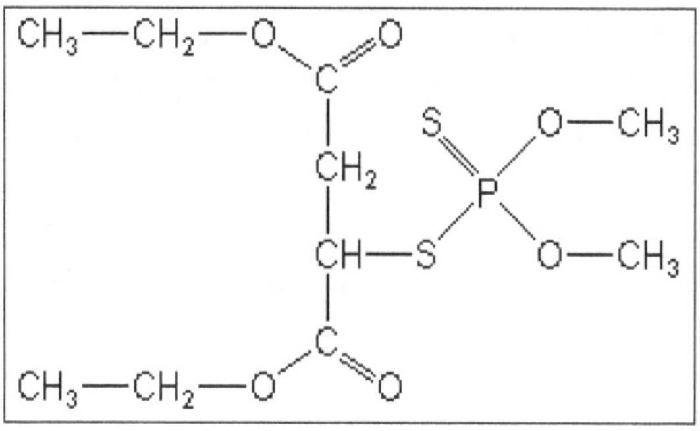

Phenyl Derivatives

The phenyl OPs contain a phenyl ring with one of the ring hydrogens displaced by attachment to the phosphorus moiety and other hydrogens frequently displaced by Cl, NO_2, CH_3, CN, or S. The phenyl OPs are generally more stable than the aliphatics, thus their residues are longer lasting. The first phenyl OP brought into agriculture was parathion (ethyl parathion) in 1947. Examples of other phenyl OPs are methyl parathion, profenofos (Curacron), sulprofos (Bolstar), isofenphos (Oftanol, Pryfon), fenitrothion (Sumithion), fenthion (Dasanit), and famphur (Cyflee. Warbex).

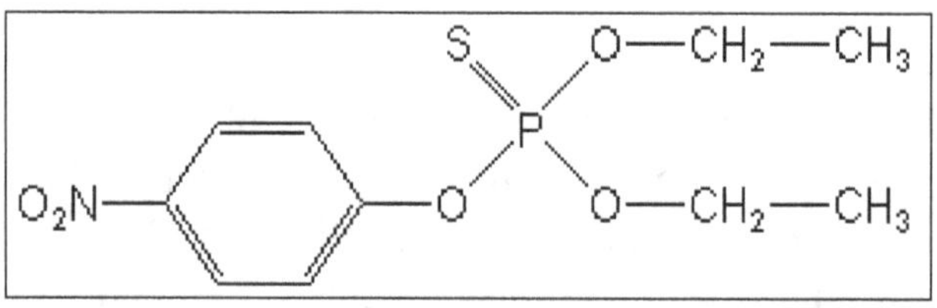

Heterocyclic Derivatives

The term heterocyclic means that the ring structures are composed of different or unlike atoms, e.g.,oxygen, nitrogen or sulfur. The first of this group was diazinon introduced in 1952. Other examples in this group are azinphos-methyl (Guthion), azinphos-ethyl (Acifon, Gusathion), chlorpyrifos (Dursban, Lorsban, Lock-On), methidathion (Supracide), phosmet (Imidan), isazophos (Brace, Triumph), and chlorpyrifos-methyl (Reldan).

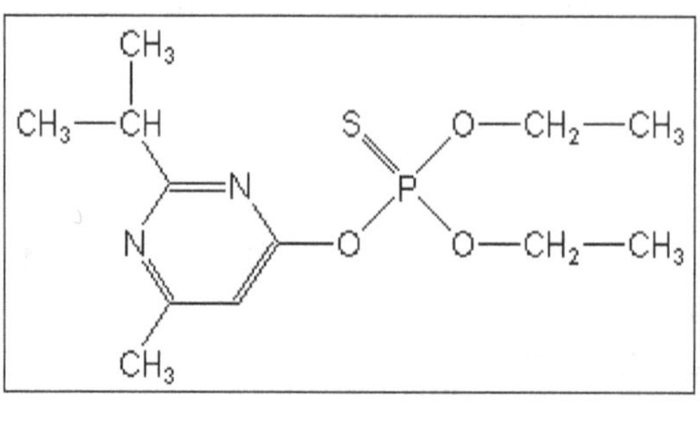

Organosulfurs

These few materials have very low toxicity to insects and are used only as acaricides (miticides). They contain two phenyl rings, resembling DDT, with sulfur in place of carbon as the central atom. These include tetradifon (Tedion), propargite (Omite, Comite), and ovex (Ovotran).

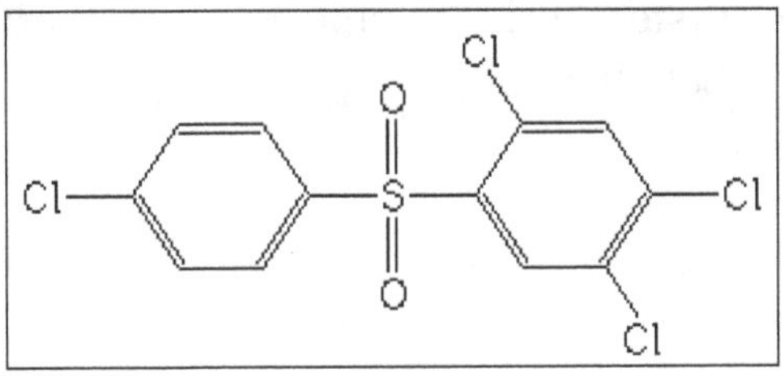

Carbamates

The carbamate insecticides are derivatives of carbamic acid (as the OPs are derivatives of phosphoric acid). And like the OPs, their mode of action is that of inhibiting the vital enzyme cholinesterase (ChE).

The first successful carbamate insecticide, carbaryl (Sevin), was introduced in 1956. More of it has been used worldwide than all the remaining carbamates combined. Two distinct qualities have made it the most popular carbamate: it's very low mammalian oral and dermal toxicity and an exceptionally broad spectrum of insect control. Other long-standing carbamate insecticides are methomyl (Lannate), carbofuran (Furadan), aldicarb (Temik), oxamyl (Vydate), thiodicarb (Larvin), methiocarb (Mesurol), propoxur (Baygon), bendiocarb (Ficam), carbosulfan (Advantage), aldoxycarb (Standak), promecarb (Carbamult), and fenoxycarb (Logic, Torus). Carbamates more recently introduced include primicarb, indoxacarb, alanycarb and furathiocarb.

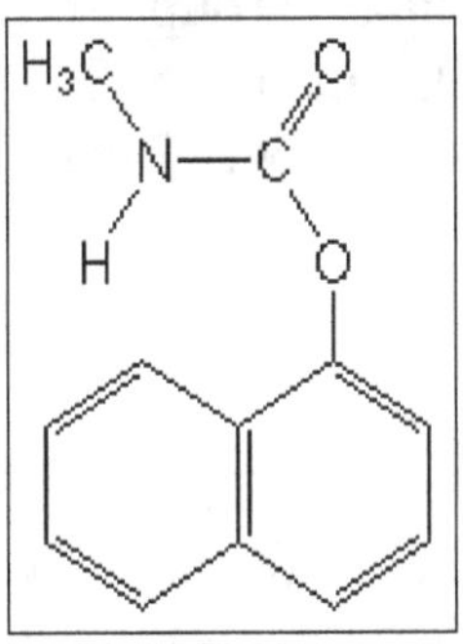

Mode of Action

Carbamates inhibit cholinesterase (ChE) as OPs do, and they behave in almost identical manner in biological systems, but with two main differences. Some carbamates are potent inhibitors of aliesterase (miscellaneous aliphatic esterases whose exact functions are not known), and their selectivity is sometimes more pronounced against the ChE of different species. Second, ChE inhibition by carbamates is reversible. When ChE is inhibited by a carbamate, it is said to be carbamylated, as when an OP results in the enzyme are phosphorylated. In insects, the effects of OPs and carbamates are primarily those of poisoning of the central nervous system, since the insect neuromuscular junction is not cholinergic, as in mammals. The only cholinergic synapses known in insects are in the central nervous system. (The chemical neuromuscular junction transmitter in insects is thought to be glutamic acid).

Formamidines

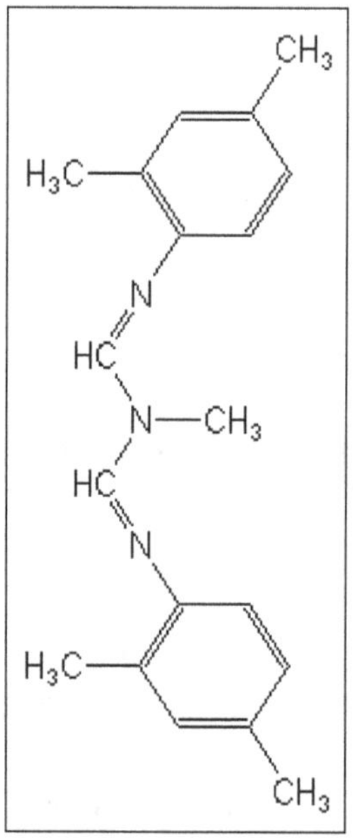

The formamidines comprise a small group of insecticides. Three examples are chlordimeform (Galecron, Fundal), which is no longer registered in the U.S., formetanate (Carzol), and amitraz (Mitac, Ovasyn). Their current value lies in the control of OP- and carbamate-resistant pests.

Mode of Action

Formamidine poisoning symptoms are distinctly different from other insecticides. Their proposed action is the inhibition of the enzyme monoamine oxidase, which is responsible for degrading the neurotransmitters norepinephrine and serotonin. This results in the accumulation of these compounds, which are known as biogenic amines. Affected insects become quiescent and die.

Dinitrophenols

The basic dinitrophenol molecule has a broad range of toxicities--as herbicides, insecticides, ovicides, and fungicides. Of the insecticides, binapacryl (Morocide) and dinocap (Karathane) were the most recently used. Dinocap is an effective miticide and was very heavily used as a fungicide for the control of powdery mildew fungi. Because of the inherent toxicity of the dinitrophenols, they have all been withdrawn.

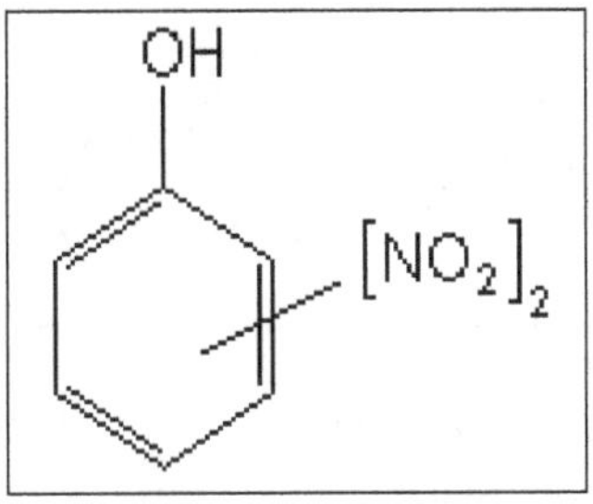

Mode of Action

Dinitrophenols act by uncoupling or inhibiting oxidative phosphorylation, which basically prevents the formation of the high-energy phosphate molecule, adenosine triphosphate (ATP).

Organotins

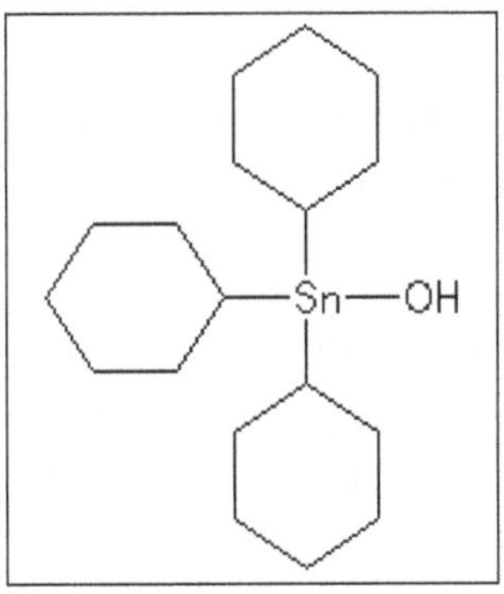

The organotins are a group of acaricides that double as fungicides. Of particular interest is cyhexatin (Plictran), one of the most selective acaricides known, introduced in 1967. Fenbutatin-oxide (Vendex) has been used extensively against mites on deciduous fruits, citrus, greenhouse crops, and ornamentals.

Mode of Action

These tin compounds inhibit oxidative phosphorylation at the site of dinitrophenol uncoupling, preventing the formation of the high-energy phosphate molecule adenosine triphosphate (ATP). These trialkyl tins also inhibit photophosphorylation in chloroplasts, the chlorophyll-bearing subcellular units) and could therefore serve as algicides.

Pyrethroids

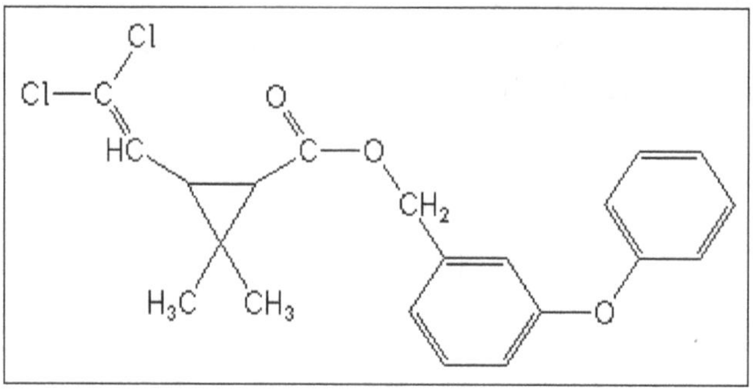

Natural pyrethrum has seldom been used for agricultural purposes because of its cost and instability in sunlight. In recent decades, many synthetic pyrethrin-like materials have become available. They were originally referred to as synthetic pyrethroids. Currently, the better nomenclature is simply pyrethroids. These are stable in sunlight and are generally effective against most agricultural insect pests when used at the very low rates of 0.01 to 0.1 pound per acre.

The pyrethroids have an interesting evolution, which is conveniently divided into four generations. The first generation contains only one pyrethroid, allethrin (Pynamin), which appeared in 1949. Its synthesis was very complex, involving 22 chemical reactions to reach the final product.

The second generation includes tetramethrin (Neo-Pynamin), followed by resmethrin (Synthrin) in 1967 (20X as effective as pyrethrum), then bioresmethrin (50X as effective as pyrethrum), then Bioallethrin, and finally phonothrin (Sumithrin).

The third generation includes fenvalerate (Pydrin [discontinued], Tribute, & Bellmark), and permethrin (Ambush, Astro, Dragnet, Flee, Pounce, Prelude, Talcord & Torpedo) which appeared in 1972-73. These became the first agricultural pyrethroids because of their exceptional insecticidal activity (0.1 lb ai/A) and their photostability. They were virtually unaffected by ultraviolet in sunlight, lasting 4-7 days as efficacious residues on crop foliage.

The fourth and current generation, is truly exciting because of their effectiveness in the range of 0.01 to 0.05 lb ai/A. These include bifenthrin (Capture, Talstar), lambda-cyhalothrin (Demand, Karate, Scimitar & Warrior), cypermethrin (Ammo, Barricade, Cymbush, Cynoff & Ripcord), cyfluthrin (Baythroid, Countdown, Cylense, Laser & Tempo), deltamethrin (Decis) esfenvalerate (Asana, Hallmark), fenpropathrin (Danitol), flucythrinate (Cybolt, Payoff), fluvalinate (Mavrik, Spur, discontinued), prallethrin (Etoc), tau-fluvalinate (Mavrik) tefluthrin (Evict, Fireban, Force & Raze), tralomethrin (Scout X-TRA, Tralex), and zeta-cypermethrin (Mustang & Fury). All of these are photostable, that is, they do not undergo photolysis (splitting) in sunlight. And because they have minimal volatility they provide extended residual effectiveness, up to 10 days under optimum conditions.

Recent additions to the fourthgeneration pyrethroids are acrinathrin (Rufast), imiprothrin (Pralle), registered in 1998, and gamma-cyhalothrim (Pytech), which is in development.

Mode of Action

The pyrethroids share similar modes of action, resembling that of DDT, and are considered axonic poisons. They apparently work by keeping open the sodium channels in neuronal membranes. There are two types of pyrethroids. Type I, among other physiological responses, have a negative temperature coefficient, resembling that of DDT. Type II, in contrast have a positive temperature coefficient, showing increased kill with increase in ambient temperature. Pyrethroids affect both the peripheral and central nervous system of the insect. They initially stimulate nerve cells to produce repetitive discharges and eventually cause paralysis. Such effects are caused by their action on the sodium channel, a tiny hole through which sodium ions are permitted to enter the axon to cause excitation. The stimulating effect of pyrethroids is much more pronounced than that of DDT.

Nicotinoids

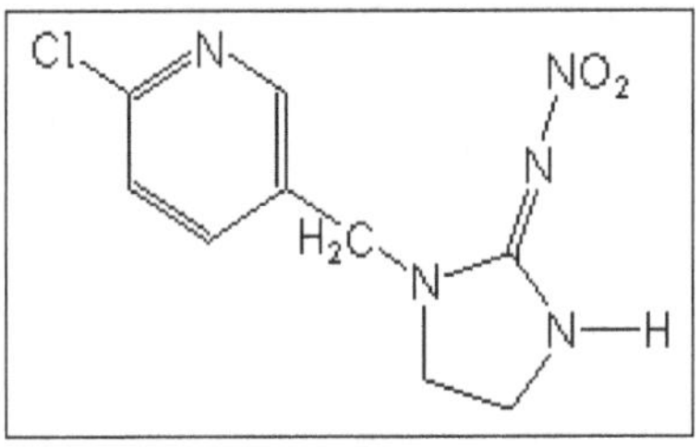

The nicotinoids are a newer class of insecticides with a new mode of action. They have been previously referred to as nitro-quanidines, neonicotinyls, neonicotinoids, chloronicotines, and more recently as the chloronicotinyls. Just as the synthetic pyrethroids are similar to and modeled after the natural pyrethrins, so too, are the nicotinoids similar to and modeled after the natural nicotine. Imidacloprid was introduced in Europe and Japan in 1990 and first registered in the U.S. in 1992. It is currently marketed as several proprietary products worldwide, e.g., Admire, Confidor, Gaucho, Merit, Premier, Premise and Provado. Very possibly it is used in the greatest volume globally of all insecticides.

Imidacloprid is a systemic insecticide, having good root-systemic characteristics and notable contact and stomach action. It is used as a soil, seed or foliar treatment in cotton, rice cereals, peanuts, potatoes, vegetables, pome fruits, pecans and turf, for the control of sucking insects, soil insects, whiteflies, termites, turf insects and the Colorado potato beetle, with long residual control. Imidacloprid has no effect on mites or nematodes.

Other nicotinoids include acetamiprid (Assail), thiamethoxam (Actara, Platinum), nitenpyram (Bestguard), clothianidin (Poncho, dinotefuran (Starke) and thiacloprid. U.S. registrations for acetamiprid, thiamethoxam and thiacloprid were granted in 2002 and clothianidin in 2003.

Mode of Action

The nicotinoids act on the central nervous system of insects, causing irreversible blockage of postsynaptic nicotinergic acetylcholine receptors.

Spinosyns

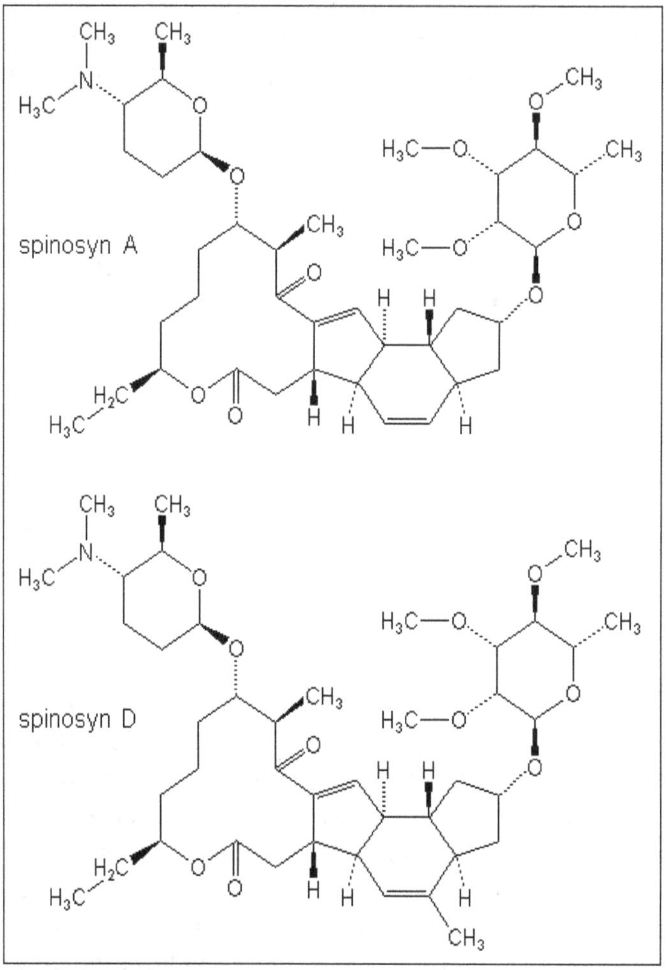

Spinosyns are among the newest classes of insecticides, represented by spinosad (Success, Tracer Naturalyte). Spinosad is a fermentation metabolite of the actinomycete Saccharopolyspora spinosa, a soil-inhabiting microorganism. It has a novel molecular structure and mode of action that provide excellent crop protection typically associated with synthetic insecticides, first registered for use on cotton in 1997. Spinosad is a mixture of spinosyns A and D (thus its name, spinosAD). It is particularly effective as a broad-spectrum material for most caterpillar pests at the astonishing rates of 0.04 to 0.09 pound of active ingredient (18 to 40 grams) per acre. It has both contact and stomach activity against lepidopteran larvae, leaf miners, thrips, and termites, with long residual activity. Crops registered include cotton, vegetables, tree fruits, ornamentals and others.

Mode of Action

Spinosad acts by disrupting binding of acetylcholine in nicotinic acetylcholine receptors at the postsynaptic cell.

Fiproles

Fipronil (Regent, Icon, Frontline) is the only insecticide in this new class, introduced in 1990 and registered in the U.S. in 1996. It is a systemic material with contact and stomach activity. Fipronil is used for the control of many soil and foliar insects, (e.g., corn rootworm, Colorado potato beetle, and rice water weevil) on a variety of crops, primarily corn, turf, and for public health insect control. It is also used for seed treatment and formulated as baits for cockroaches, ants and termites. Fipronil is effective against insects resistant or tolerant to pyrethroid, organophosphate and carbamate insecticides.

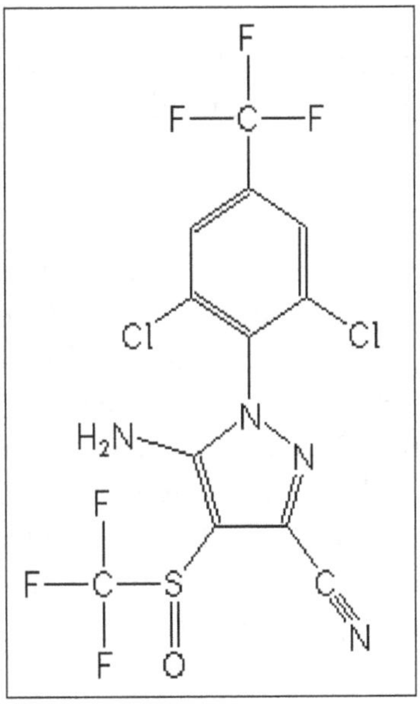

Mode of Action

Fipronil blocks the (g-aminobutyric acid- (GABA) regulated chloride channel in neurons, thus antagonizing the "calming" effects of GABA, similar to the action of the Cyclodienes.

Pyrroles

Chlorfenapyr (Alert, Pirate) is the first and only member of this unique chemical group, as both a contact and stomach insecticide-miticide. It is used on cotton and experimentally on corn, soybeans, vegetables, tree and vine crops, and ornamentals to control whitefly, thrips, caterpillars, mites, leafminers, aphids, and Colorado potato beetle. It has ovicidal activity on some species. EPA took the unusual step of refusing to register chlorfenapyr in 2000 for cotton insect control because of potential hazards to birds. However, labels for greenhouse ornamentals were granted in 2001.

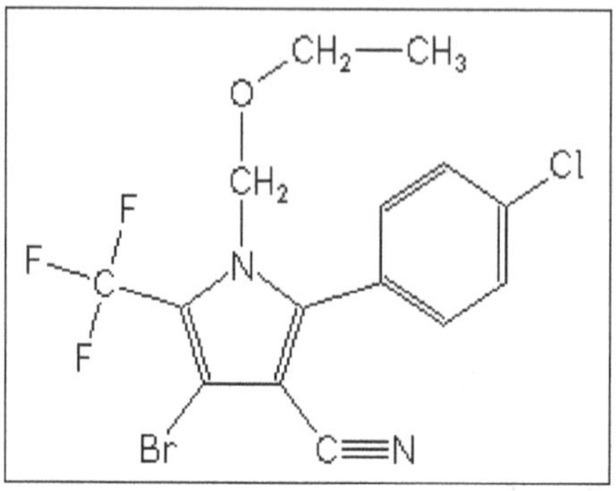

Mode of Action

Chlorfenapyr is an "uncoupler" or inhibitor of oxidative phosphorylation, preventing the formation of the crucial energy molecule adenosine triphosphate (ATP).

Pyrazoles

The original pyrazoles were tebufenpyrad and fenpyroximate (not illustrated). These were designed primarily as non-systemic contact and stomach miticides, but do have limited effectiveness on psylla, aphids, whitefly, and thrips. Tebufenpyrad (Pyranica, Masai), registered by EPA in 2002, is used on cotton, soybeans, vegetables, pome fruits, grapes and citrus. Fenpyroximate (Acaban, Dynamite) controls all stages of mites, gives fast knockdown, inhibits molting of immature stages of mites, and has long residual activity. Newer members of this class include ethiprole (Curbex) which is active on a broad sprectum of chewing and sucking insects, and tolfenpyrad (OMI-88) which is reputed to active on pests infesting cole and cucurbit crops.

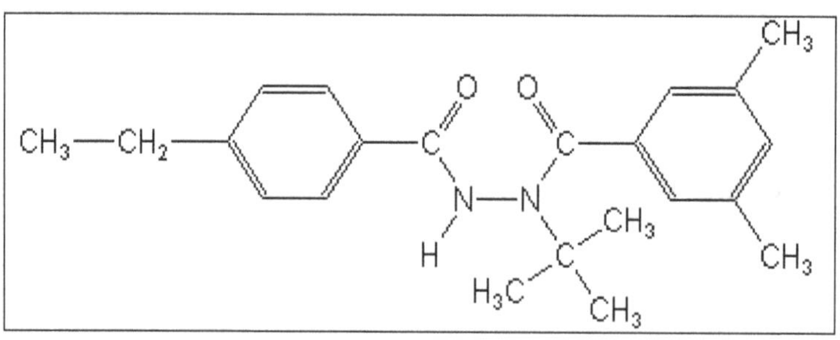

Mode of Action

Their mode of action is that of inhibiting mitochondrial electron transport at the NADH-CoQ reductase site, leading to the disruption of adenosine triphosphate (ATP) formation, the crucial energy molecule.

Pyridazinones

Pyridaben (Nexter, Sanmite) is the only member of this class. It is a selective contact insecticide and miticide, also effective against thrips, aphids, whiteflies and leafhopprs. Registrations are for pome fruits, almonds, citrus, ornamentals and greenhouse ornamentals. Pyridaben provides exceptionally long residual control, and rapid knockdown at a broad range of temperatures.

Mode of Action

Pyridaben is a metabolic inhibitor that interrupts mitochondrial electron transport at Site 1, similar to the Quinazolines, below.

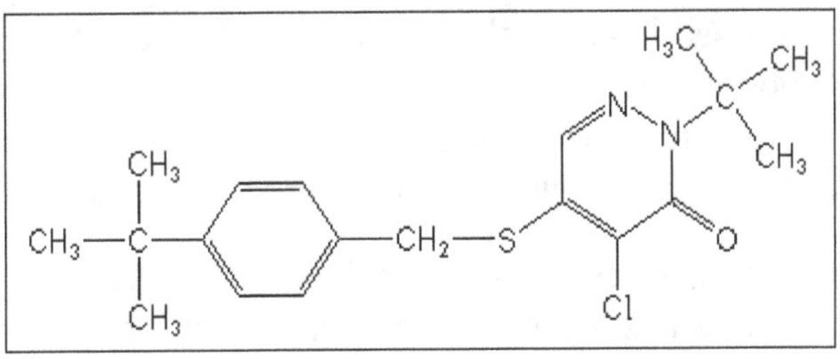

Quinazolines

The quinazolines offer a unique chemical configuration, consisting only of one insecticide, fenazaquin (Matador). Fenazaquin is a contact and stomach miticide. It has ovicidal activity, gives rapid knockdown, and controls all stages of mites. Not yet registered in the U.S., it is used on cotton, stone and pome fruits, citrus, grapes and ornamentals.

Fenazaquin

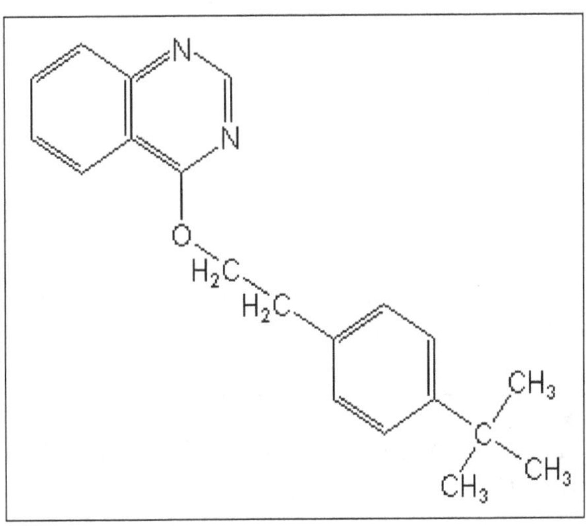

Mode of Action

Fenazaquin inhibits mitochondrial electron transport at Site 1, similar to the Pyridazinones, above.

Benzoylureas

Benzoylureas are an entirely different class of insecticides that act as insect growth regulators (IGRs). Rather than being the typical poisons that attack the insect nervous system, they interfere with chitin synthesis and are taken up more by ingestion than by contact. Their greatest value is in the control of caterpillars and beetle larvae.

Benzoylureas were first used in Central America in 1985, to control a severe, resistant leafworm complex (Spodoptera spp., Trichoplusia spp.) outbreak in cotton. The withdrawal of the ovicide chlordimeform made their control quite difficult due to their high resistance to almost all insecticide classes, including the pyrethroids.

The benzoylureas were introduced in 1978 by Bayer of Germany, triflumuron (Alsystin) being the first. Others appearing since then are chlorfluazuron (Atabron), followed by teflubenzuron (Nomolt, Dart), hexaflumuron (Trueno, Consult), flufenoxuron (Cascade), and flucycloxuron (Andalin). Others are flurazuron, novaluron, and diafenthiuron, bistrifluron (DBI-3204) and noviflumuron (XDE-007). Until recently lufenuron (Axor) was the newest addition to this group, appearing in 1990. Among the newer benzoylureas only hexaflumuron and novaluron have been registered by EPA.

The only other benzoylurea registered in the U.S. is diflubenzuron (Dimilin, Adept, Micromite). It was first registered in 1982 for gypsy moth, cotton boll weevil, most forest caterpillars, soybean caterpillars, and mushroom flies, but now with a much broader range of registrations.

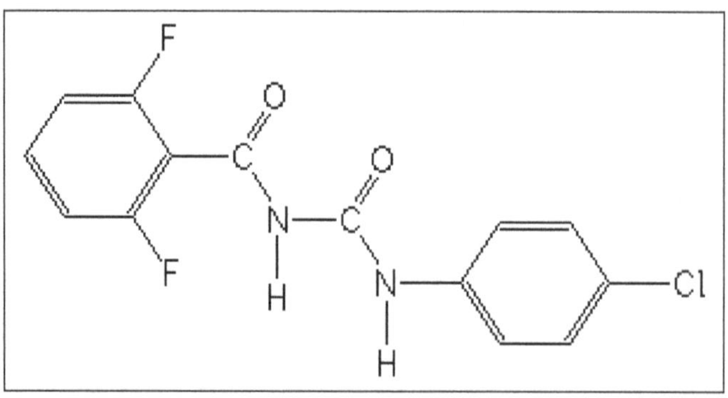

Though not a benzoylurea, cyromazine (Larvadex, Trigard), a triazine, is also a potent chitin synthesis inhibitor. It is selective toward Dipterous species and used for the control of leafminers in vegetable crops and ornamentals, and fed to poultry or sprayed to control flies in manure of broiler and egg producing operations, and incorporated into compost of mushroom houses for fungus gnats.

Mode of Action

The benzoylureas act on the larval stages of most insects by inhibiting or blocking the synthesis of chitin, a vital and almost indestructible part of the insect exoskeleton. Typical effects on developing larvae are the rupture of malformed cuticle or death by starvation. Adult female boll weevils exposed to diflubenzuron lay eggs that do not hatch. And, mosquito larvae control can be achieved with as little as 1.0 gram of diflubenzuron per acre of surface water.

Botanicals

Botanical insecticides are of great interest to many, for they are natural insecticides, toxicants derived from plants. Historically, the plant materials have been in use longer than any other group, with the possible exception of sulfur. Tobacco, pyrethrum, derris, hellebore, quassia, camphor, and turpentine were some of the more important plant products in use before the organized search for insecticides began in the early 1940s.

In recent years the term biorational has been put into play by the EPA. There are similarities and differences between the terms botanical and biorational. Botanical insecticide use in the U.S. peaked in 1966, and has declined steadily since. Pyrethrum is now the only classical botanical of significance in use. Some newer plant-derived insecticides that have come into use are referred to as florals or scented plant chemicals and include, among others, limonene, cinnamaldehyde and eugenol. In addition, there is azadirachtin from the neem tree which is used in greenhouse and on ornamentals.

Pyrethrum

Pyrethrum is extracted from the flowers of a chrysanthemum grown in Kenya and Ecuador. It is one of the oldest and safest insecticides available. The ground, dried flowers were used in the early 19th century as the original louse powder to control body lice in the Napoleonic Wars. Pyrethrum acts on insects with phenomenal speed causing immediate paralysis, thus its popularity in fast knockdown household aerosols. However, unless it is formulated with one of the synergists, most of the paralyzed insects recover to once again become pests. Pyrethrum is a mixture of four compounds: pyrethrins I and II and cinerins I and II.

Mode of Action

Pyrethrum is an axonic poison, as are the synthetic pyrethroids and DDT. Axonic poisons are those that in some way affect the electrical impulse transmission along the axons, the elongated extensions of the neuron cell body. Pyrethrum and some pyrethroids have a greater insecticidal effect when the temperature is lowered, a negative temperature coefficient, as does DDT. They affect both the peripheral and central nervous system of the insect. Pyrethrum initially stimulates nerve cells to produce repetitive discharges, leading eventually to paralysis. Such effects are caused by their action on the sodium channel, a tiny hole through which sodium ions are permitted to enter the axon to cause excitation. These effects are produced in insect nerve cord, which contains ganglia and synapses, as well as in giant nerve fiber axons.

Nicotine

Nicotine is extracted by several methods from tobacco, and is effective against most all types of insect pests, but is used particularly for aphids and caterpillars--soft bodied insects. Nicotine is an alkaloid, a chemical class of heterocyclic compounds containing nitrogen and having prominent physiological properties. Other well-known alkaloids that are not insecticides are caffeine (coffee, tea), quinine (cinchona bark), morphine (opium poppy), cocaine (coca leaves), ricinine (a poison in castor oil beans), strychnine (Strychnos nux vomica), coniine (spotted hemlock, the poison used by Socrates), and, finally LSD (a hallucinogen from the ergot fungus attacking grain).

Mode of Action

Nicotine action is one of the first, classic modes of action identified by pharmacologists. Drugs that act similarly to nicotine are said to have a nicotinic response. Nicotine mimics acetylcholine (ACh) at the neuromuscular (nerve/muscle) junction in mammals, and results in twitching, convulsions, and death, all in rapid order. In insects the same action is observed, but only in the central nervous system ganglia.

Rotenone

Rotenone or rotenoids are produced in the roots of two genera of the legume family: Derris and Lonchocarpus (also called cubé) grown in South America. It is both a stomach and contact insecticide and used for the last century and a half to control leaf-eating caterpillars, and three centuries prior to that in South America to paralyze fish, causing them to surface and be easily captured. Today, rotenone is used in the same way to reclaim lakes for game fishing. Used on a prescribed basis, it eliminates all fish, closing the lake to reintroduction of rough species. It is a selective pesticide in that it kills all fish at dosages that are relatively nontoxic to fish food organisms, and is degraded rapidly.

Mode of Action

Rotenone is a respiratory enzyme inhibitor, acting between NAD+ (a coenzyme involved in oxidation and reduction in metabolic pathways) and coenzyme Q (a respiratory enzyme responsible for carrying electrons in some electron transport chains), resulting in failure of the respiratory functions.

Limonene or D-Limonene

Limonene or d-Limonene is the latest addition to the botanicals. Limonene belongs to a group often called floral or scented plant chemicals. Extracted from citrus peel, it is effective against all external pests of pets, including fleas, lice, mites, and ticks, and is virtually nontoxic to warm-blooded animals. Several insecticidal substances occur in citrus oil, but the most important is limonene, which constitutes about 98% of the orange peel oil by weight. Two other recently introduced floral products are eugenol (Oil of Cloves) and cinnamaldehyde (derived from Ceylon and Chinese cinnamon oils). They are used on ornamentals and many crops to control various insects.

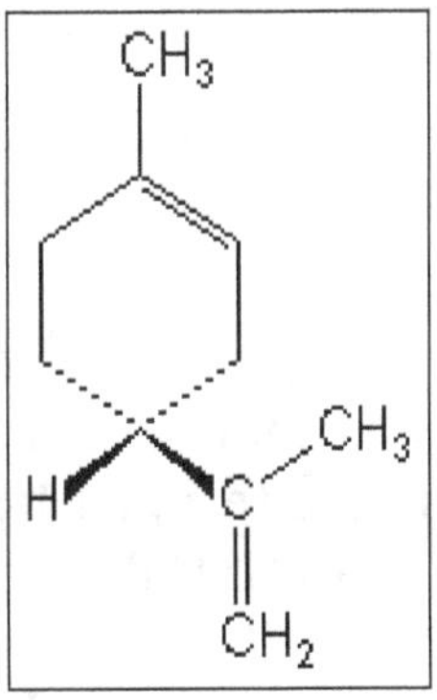

Mode of Action

Its mode of action is similar to that of pyrethrum. It affects the sensory nerves of the peripheral nervous system, but it is not a ChE inhibitor.

Neem

Neem oil extracts are squeezed from the seeds of the neem tree and contain the active ingredient azadirachtin, a nortriterpenoid belonging to the lemonoids. Azadirachtin has shown some rather sensational insecticidal, fungicidal and bactericidal properties, including insect growth regulating qualities. Azatin is marketed as an insect growth regulator, and Align and Nemix as a stomach/contact insecticide for greenhouse and ornamentals.

Mode of Action

Azadirachtin disrupts molting by inhibiting biosynthesis or metabolism of ecdysone, the juvenile molting hormone.

Synergists or Activators

Synergists are not in themselves considered toxic or insecticidal, but are materials used with insecticides to synergize or enhance the activity of the insecticides. The first was introduced in 1940 to increase the effectiveness of pyrethrum. Since then many materials have appeared, but only a few are still marketed. Synergists are found in most all household, livestock and pet aerosols to enhance the action of the fast knockdown insecticides pyrethrum, allethrin, and resmethrin, against flying insects. Current synergists, such as piperonyl butoxide, contain the methylenedioxyphenyl moiety, a molecule found in sesame oil and later named sesamin.

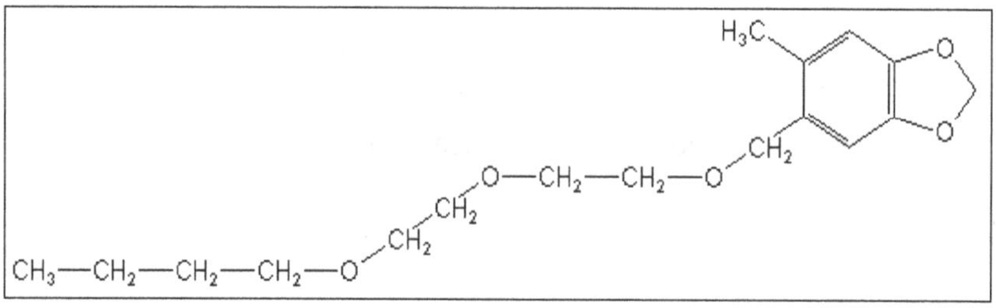

Mode of Action

The synergists inhibit cytochrome P-450 dependent polysubstrate monooxygenases (PSMOs), enzymes produced by microsomes, the subcellular units found in the liver of mammals and in some insect tissues (e.g., fat bodies). The earlier name for these enzymes was mixed-function oxidases (MFOs). These PSMOs bind the enzymes that degrade selected foreign substances, such as pyrethrum, allethrin, resmethrin or any other synergized compound. Synergists simply bind the oxidative enzymes and prevent them from degrading the toxicant.

Antibiotics

In this category belong the avermectins, which are insecticidal, acaricidal, and antihelminthic agents that have been isolated from the fermentation products of Streptomyces avermitilis, a member of the actinomycete family. Abamectin is the common name assigned to the avermectins, a mixture of containing 80% avermectin B1a and 20% B1b, homologs that have about equal biological activity. Clinch is fire ant bait, and Avid is applied as a miticide/insecticide. Abamectin has certain local systemic qualities, permitting it to kill mites on a leaf's underside when only the upper surface is treated. The most promising uses for these materials are the control of spider mites, leafminers and other difficult-to-control greenhouse pests, and internal parasites of domestic animals.

Emamectin benzoate (Proclaim, Denim) is an analog of abamectin, produced by the same fermentation system as abamectin. It was first registered in 1999. It is both a stomach and contact insecticide used primarily for control of caterpillars at the rate of

0.0075 to 0.015 lb (3.5 to 7.0 grams) a.i. per acre. Shortly after exposure, larvae stop feeding and become irreversibly paralyzed, dying in 3-4 days. Rapid photodegradation of both abamectin and emamectin occurs on the leaf surface. More recently, milabectin (Mesa) has been introduced. It is a miticide with activity on piercing/sucking insects and is pending for registration.

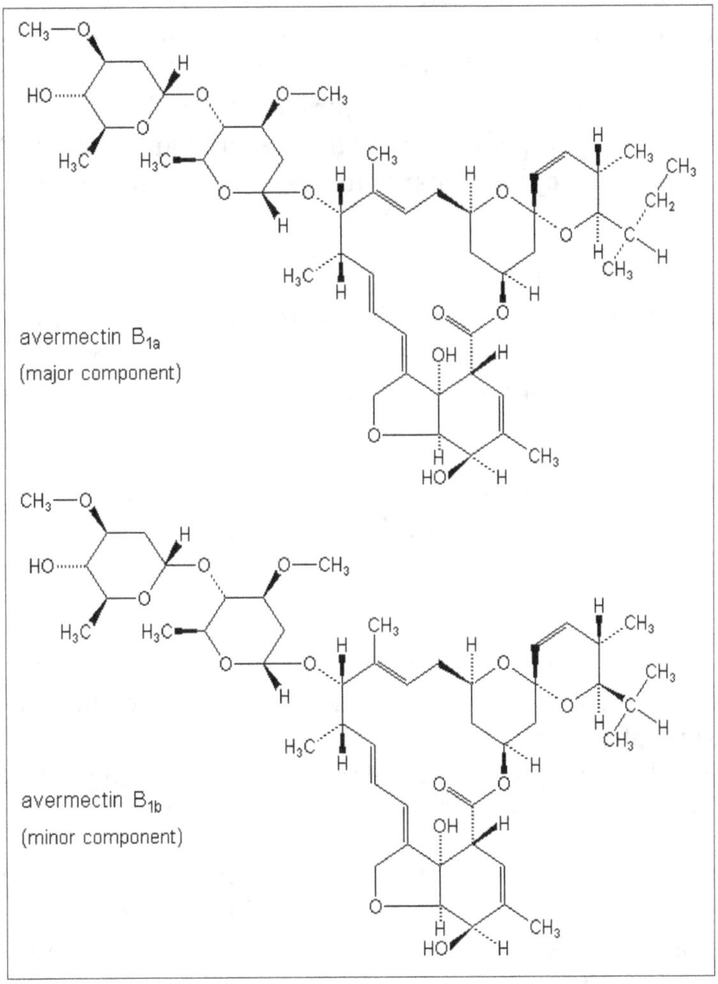

avermectin B_{1a} (major component)

avermectin B_{1b} (minor component)

Mode of Action

Avermectins block the neurotransmitter (g-aminobutyric acid (GABA) at the neuro-muscular junction in insects and mites. Visible activity, such as feeding and egg laying, stops shortly after exposure, though death may not occur for several days.

Fumigants

The fumigants are small, volatile, organic molecules that become gases at temperatures above 40 °F. They are usually heavier than air and commonly contain one or more of the halogens (Cl, Br, or F). Most are highly penetrating, reaching through large masses of material. They are used to kill insects, insect eggs, nematodes, and certain microorganisms in buildings, warehouses, grain elevators, soils, and greenhouses and in packaged products such as dried fruits, beans, grain, and breakfast cereals.

Although its use is now in decline because of environmental concerns, methyl bromide is the most heavily used of the fumigants, 68,424 metric tons worldwide in 1996, almost half of which is used in the U.S. The dominant use is for preplanting soil treatments, which accounted for 70% of that global total. Quarantine uses account for 5-8%, while 8% is used to treat perishable products, such as flowers and fruits, and 12% for nonperishable products, like nuts and timber. Approximately 6% is used for structural applications, as in drywood termite fumigation of infested buildings.

With the recently passed change to the Clean Air Act amendments of 1990, U.S. production and importation must be reduced 25% from 1991 levels by 1999. A 50% reduction must be achieved by 2001, followed by a 70% reduction in 2003, and a full ban of the product in 2005. Under the Montreal Protocol, developing countries have until 2015 to phase out methyl bromide production.

Some of the other common fumigants are ethylene dichloride, hydrogen cyanide, sulfuryl fluoride (Vikane), Vapam, TeloneII, D-D, chlorothene, ethylene oxide, and the familiar home-use moth repellents napthalene crystals and paradichlorobenzene crystals.

Phosphine gas (PH_3) has also replaced methyl bromide in a few applications, primarily for insect pests of grain and food commodities. Treatment requires the use of aluminum or magnesium phosphide pellets, which react with atmospheric moisture to produce the gas. Phosphine, however, is very damaging to fresh commodities and is highly adsorbed onto oil, thus does not perform as a soil fumigant.

Alternatives that can fully replace methyl bromide are unlikely to be available by the deadlines set for replacement. Its low cost and utility on a wide variety of pests are hard to match. Because the loss of MeBr has considerable economic consequences, EPA has made it a priority to find and register replacements. To this end some progress has been made. The chemical 1,3-dichloropropane (Telone) was registered in 2001 for preplant soil fumigation in strawberries and tomatoes. Moreover, iodomethane (Midas) and metam-potassium (Curtin) are both being evaluated as soil fumigants.

Mode of Action

Fumigants, as a group, are narcotics. That is, they act through means more physical than physiological. The fumigants are liposoluble (fat soluble); they have common symptomology; their effects are reversible; and their activity is altered very little by structural changes in their molecules. As narcotics, they induce narcosis, sleep, or unconsciousness, which in effect is their action on insects. Liposolubility appears to be an important factor in their action, since these narcotics lodge in lipid-containing tissues found throughout the insect body, including their nervous system.

Insect Repellents

Historically, repellents have included smoke, plants hung in dwellings or rubbed on the skin as the fresh plant or its brews, oils, pitches, tars, and various earths applied to the body. Before a more edified approach to insect olfaction and behavior was developed, it was wrongly assumed that if a substance was repugnant to humans it would likewise be repellent to annoying insects.

In recent history, the repellents have been dimethyl phthalate, Indalone, Rutgers 612, dibutyl phthalate, various MGK repellents, benzyl benzoate, the military clothing repellent (N-butyl acetanilide), dimethyl carbate (Dimelone) and diethyl toluamide (DEET, Delphene). Of these, only DEET has survived, and is used worldwide for biting flies and mosquitoes. Most of the others have lost their registrations and are no longer available.

In 1999, EPA has registered a new insect repellent, N-methylneodecanamide. Rather than being used on humans to repel insects, it is applied to household floors and other surfaces to repel cockroaches and ants.

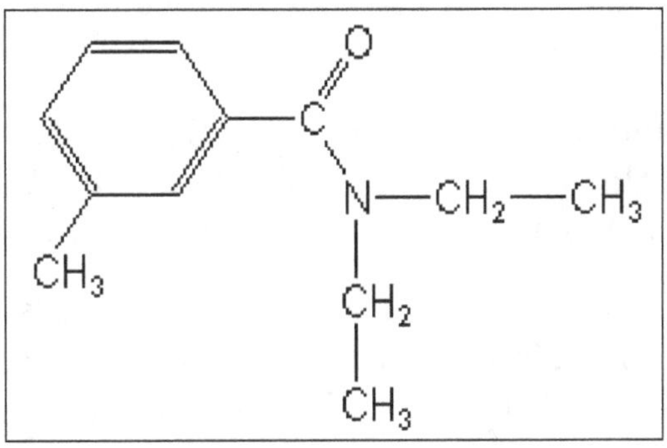

Inorganics

Inorganic insecticides are those that do not contain carbon. Usually they are white crystals in their natural state, resembling the salts. They are stable chemicals, do not evaporate, and are usually water soluble.

Sulfur, mentioned in the introduction, is very likely the oldest known, effective insecticide. Sulfur and sulfur candles were burned by our great-grandparents for every conceivable purpose, from bedbug fumigation to the cleansing of a house just removed from quarantine of smallpox. Today, sulfur is a highly useful material in integrated pest management programs where target pests specificity is important. Sulfur dusts are especially toxic to mites of every variety, such as chiggers and spider mites, and to thrips and newly-hatched scale insects. Sulfur dusts and sprays are also fungicidal, particularly against powdery mildews.

Several other inorganic compounds have been used as insecticides: mercury, boron, thallium, arsenic, antimony, selenium, and fluoride. Arsenicals have included the copper arsenate, Paris green, lead arsenate, and calcium arsenate. The arsenicals uncouple oxidative phosphorylation, inhibit certain enzymes that contain sulfhydryl (-SH) groups, and coagulate protein by causing the shape or configuration of proteins to change.

The inorganic fluorides were sodium fluoride, barium fluosilicate, sodium silicofluoride, and cryolite (Kryocide). Cryolite has returned in recent years as a relatively safe fruit and vegetable insecticide, used in integrated pest management programs. The fluoride ion inhibits many enzymes that contain iron, calcium, and magnesium. Several of these enzymes are involved in energy production in cells, as in the case of phosphatases and phosphorylases.

Boric acid, used against cockroaches and other crawling household pests in the 1930's and '40's, has also returned. As a salt, it is non-volatile and will remain effective as long as it is kept dry and in adequate concentration. Consequently, it has the longest residual activity of any insecticide used for crawling household insects, and is quite useful in the control of all cockroach species when placed in wall voids and other protected, difficult-to-reach sites. It acts as a stomach poison and insect cuticle wax absorber.

Sodium borate (disodium octaborate tetrahydrate) (Tim-Bor, Bora-Care) resembles boric acid in its action. This water-soluble salt is used to treat lumber and other wood products to control decay fungi, termites, and other wood infesting pests.

The last group of inorganics is the silica gels or silica aerogels-light, white, fluffy, silicate dusts used for household insect control. The silica aerogels kill insects by absorbing waxes from the insect cuticle, permitting the continuous loss of water from the insect body, causing the insects to become desiccated and die from dehydration. These include Dri-Die, Drianone, and Silikil Microcel. Drianone is fortified with pyrethrum and synergists to enhance its effectiveness.

New Miscellaneous Insecticide Classes

Seven classes of insecticides have made their appearance in recent years. These are summarized below:

Methoxyacrylates

Fluacrypyrin (Titaron) is an acaricide for fruit and is the only example of this class currently. It is registered for use on fruit in Japan.

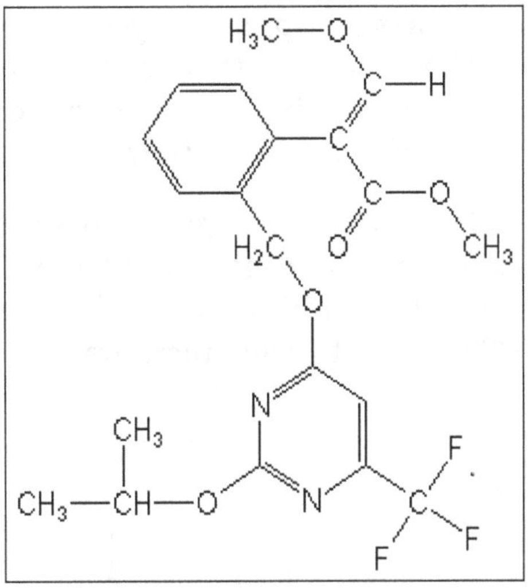

Fluacrypyrim (titaron): methyl (E)-2-{α-[2-isopropoxy-6-(trifluoromethyl) pyrimidin-4-yloxy]-o-tolyl}-3-methoxyacrylate.

Naphthoquinones

Acequinocyl (Kanemite, Piton) is a miticide with insecticidal activity for pome fruit, nut crops, citrus and ornamentals. It is the only member of this group at present and the mode of action is not yet determined. It holds registrations in Korea and Japan but not in the US.

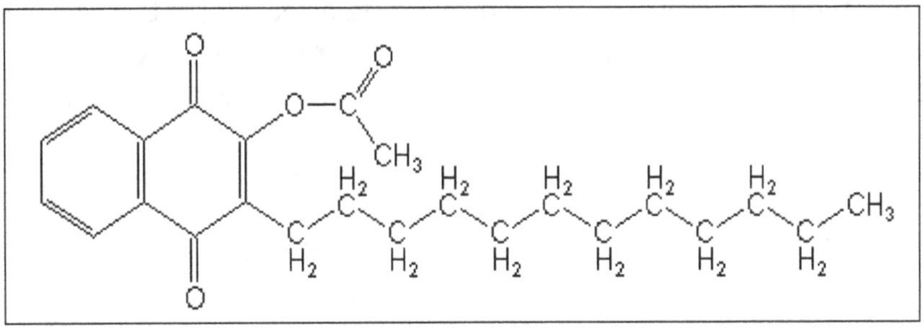

Acequinocyl (Kanemite, Piton): 3-dodecyl-1,4-dihydro-1,4-dioxo-2-naphthyl acetate.

Nereistoxin Analogues

These include thiocyclam, cartap, bensultap, and thiocytap-sodium. Analogues of nereistoxin have been known for decades. They generally are stomach poisons with some contact action and often show some systemic action. A major share of the development and use of these compounds has taken place in Japan. They are based on a natural toxin of the marine worm Lumbriconereis heteropoda. Of the many analogs synthesized only those that were metabolized back to the original nereistoxin after application were active. In this sense members of this class are proinsecticides in that they are applied in their manufactured form but are known to degrade to a specific active component. The members of this group tend to be selectively active on Colopteran and Lepidopteran insect pests. Cartap (Agrotap) is a broad spectrum insecticide with good activity against rice stem borer. Bensultap (Bancol) is used to control the Colorado Potato beetle and other insect pests. Thiosultap-sodium (Pilarhope) is used to control selected beetle and Lepidopteran pests on rice, vegetables and fruit trees.

Thiocyclam (Evisect), is used for the control of similar pests in several crops. Members of this class act as acetyl choline receptor agonists at low concentrations and as channel blockers at higher concentrations. Although there has been commercial interest in thiocyclam for use in the US we do not believe there are commercial examples that are to achieve U.S. registration.

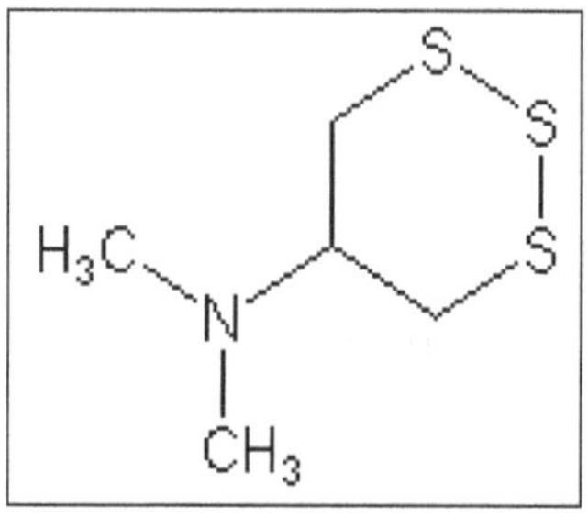

Thiocyclam (evisect): N,N-dimethyl-1,2,3-trithian-5-ylamine.

Pyridine Azomethin

Pymetrozine (Fulfill), first registered in 1999 by EPA has a unique mode of action that is not fully understood. It appears to act by preventing insects from the Order Homoptera from inserting their stylus into plant tissue. Pymetrozine is used to control aphids and whiteflies in vegetables, potatoes, tobacco, deciduous citrus fruit hops and ornamentals.

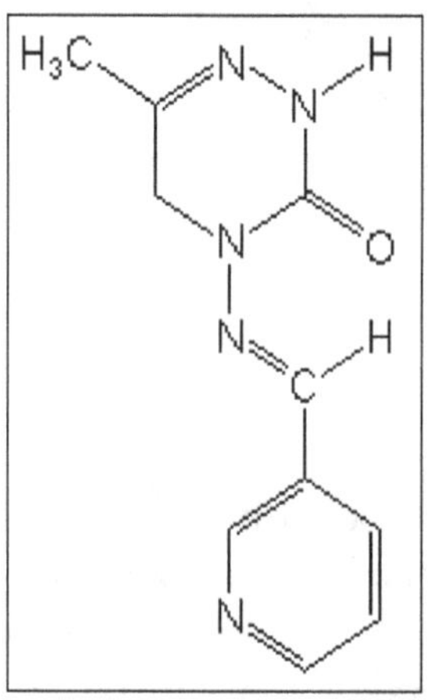

Pymetrozine (fulfill): (E)-4,5-dihydro-6-methyl-4-(3-pyridylmethyleneamino) -1,2,4-triazin-3(2H)-one.

Pyrimidinamines

Pyrimidifen (Miteclean) is an insecticide and miticide. As a miticide the product controls spider and rust mites in deciduous fruits, citrus, vegetables and tea. As an insecticide it controls diamondback moth in vegetables. Very little information is available on the other member of this class, Flufenerim, other than it is insecticidal.

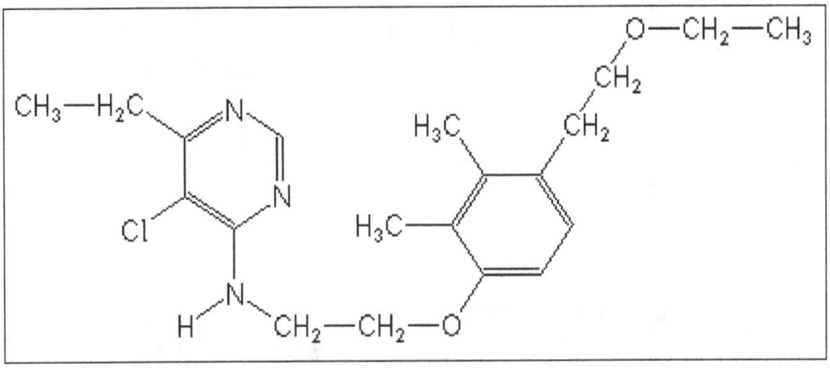

Pyrimidifen (miteclean): 5-chloro-N-{2-[4-(2-ethoxyethyl)-2, 3- dimethylphenoxy]ethyl}-6-ethylpyrimidin-4-amine.

Tetronic Acids

Spirodiclofen (Envidor) and spiromesifen (Oberon) are the only two members of this recently introduced class. Spirodiclofen has broad-spectrum activity against mites, and controls scale crawlers and psyllad nymphs. Action is good on eggs and quiescent stages. Target crops are citrus, grapes, nuts, pome and stone fruits.

Miscellaneous Compounds

Clofentezine (Apollo, Acaristop), belongs to the unique group, the tetrazines, used as an acaricide/ovicide for deciduous fruits, citrus, cotton, cucurbits, vines and ornamentals. A newer and somewhat similar agent is etoxazole (TerraSan) which is an acaricide registered in 2002 for ornamentals grown in greenhouses. The modes of action of these two compounds are not yet understood.

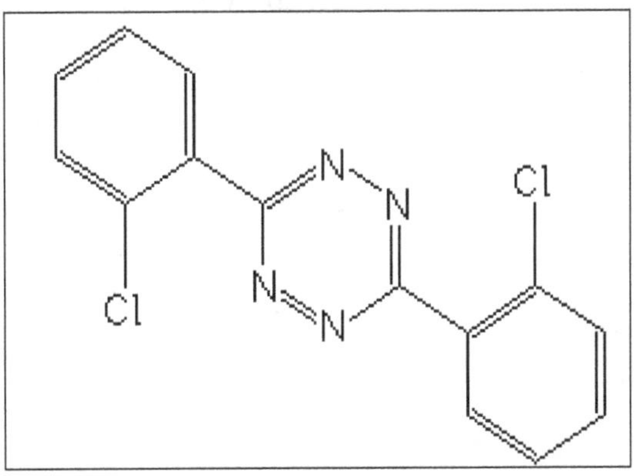

Enzone, sodium tetrathiocarbonate, is used only on grapes and citrus applied as a water application and irrigated into the soil. It breaks down in the soil to form carbon disulfide, which acts rapidly, decomposes quickly, and is effective against nematodes, soil insects, and soil borne diseases.

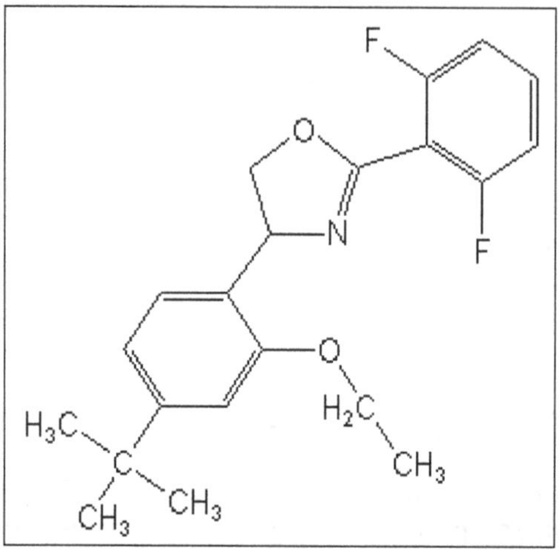

Etoxazole (terrasan): (RS)-5-tert-butyl-2-[2-(2,6-difluorophenyl)-4, 5-dihydro-1,3-oxazol-4-yl]phenetole.

The newest agents in this category are pyridanyl and amidoflumet. Pryidalyl is active on Lepidoptera and thrips and has the advantage of being active against pyrethroid-resistant insects. Little more is available on amidoflumet other than it is an acaricide in the early stage of its development.

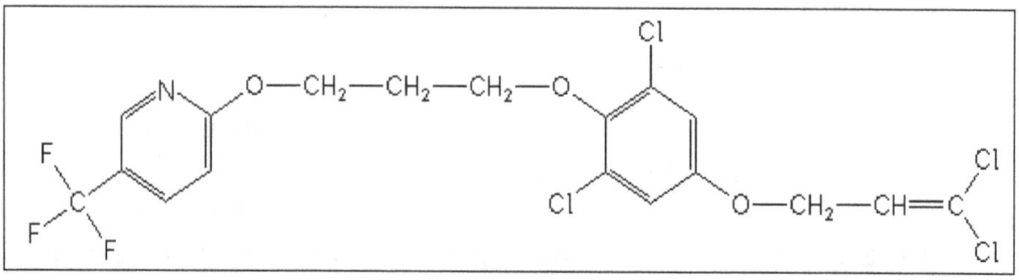

Pyridalyl: 2,6-dichloro-4-(3,3-dichloroallyloxy)phenyl 3-[5-(trifluoromethyl)-2-pyridyloxy]propyl ether.

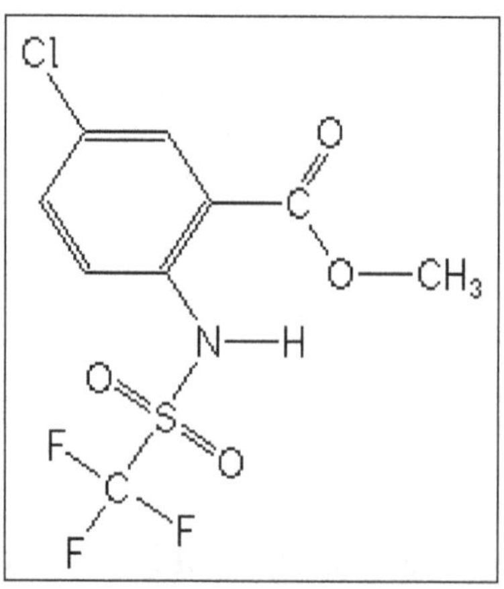

Amidoflumet: Methyl 5-chloro-2-{[(trifluoromethyl)sulfonyl]amino}benzoate.

Biorational Insecticides

The U.S. EPA identifies biorational pesticides as inherently different from conventional pesticides, having fundamentally different modes of action, and consequently, lower risks of adverse effects from their use. Biorational has come to mean any substance of natural origin (or man-made substances resembling those of natural origin), that has a detrimental or lethal effect on specific target pest(s), e.g., insects, weeds, plant diseases (including nematodes), and vertebrate pests, possess a unique mode of action, are non-toxic to man and his domestic plants and animals, and have little or no adverse effects on wildlife and the environment. EPA uses a similar term, biopesticides, which will be defined below.

Biorational insecticides are grouped as either (1) biochemicals (hormones, enzymes, pheromones and natural agents, such as insect and plant growth regulators), or (2) microbial (viruses, bacteria, fungi, protozoa, and nematodes). In the 1990's the US-EPA began to emphasize a class of products known as biopesticides. EPA places biopesticides into three categories:

- Microbial pesticides (bacteria, fungi, virus or protozoa).
- Biochemicals: Natural substances that control pests by non-toxic mechanisms. An example is insect pheromones.
- Plant-Incorporated protectants (PIPs): (Primarily transgenic plants, e.g., Bt corn).

EPA discloses that at the end of 2001 there were nearly 200 biopesticide active ingredients registered comprising nearly 800 products. Characteristics that distinguish biorational and biopesticides from conventional ones include: very low orders of toxicity to non-target species, pest targets are specific, generally low use rates, rapid decomposition in the environment, usually work well in IPM programs and reduce reliance on conventional pesticide products.

The terms "biorational" and "biopesticide" overlap but are not identical. Below is an overview of what is considered as biorational insecticides. In some cases there are overlaps with botanicals (e.g., rotenone, florals, etc. and also conventional insecticides (e.g., benzoylureas). We will point out the discrepancies in classification between the biorational and biopesticide categories where they occur.

Insect Pheromones

Most insects appear to communicate by releasing molecular quantities of highly specific compounds that vaporize readily and are detected by insects of the same species. These delicate molecules are known as pheromones. The word pheromone comes from the Greek pherein, "to carry," and hormon, "to excite or stimulate."

Of 1,314 species of insects with confirmed attraction responses to identified pheromones, 1,260 of these pheromones are produced by females. Only 54 species use male-produced sex attractants. In a few species both sexes produce the same attractant by both sexes.

Pheromones are classified as either, releasers and or primers. Releasers are fast-acting and are used by insects for sexual attraction, aggregation (including trail following), dispersion, oviposition, and alarm. Primers are slow-acting and cause gradual changes in growth and development, especially in social insects by regulating caste ratios of the colony.

The five principal uses for sex pheromones are: (1) Male trapping, to reduce the reproductive potential of an insect population; (2) Movement studies, to determine how far and where insects move from a given point; (3) Population monitoring, to determine when peak emergence or appearance occurs; (4) Detection programs, to determine if a pest occurs in a limited trapping area, such as around international airports or quarantined areas; and (5) The "confusion or mating disruption" technique.

The first use of mating disruption involved gossyplure, the pink bollworm pheromone. Incorporated into small, hollow, polyvinyl fibers that permit slow release of the pheromone, it was broadcast heavily and uniformly over infested cotton fields. In mid-2002, EPA had registered 36 pheromones which comprised over 200 individual products.

Despite praise for the potential of sex pheromones, they are most practically used in survey traps to provide information about population levels, to delineate infestations, to monitor control or eradication programs, and to warn of new pest introductions.

Fungicides

A fungicide is a specific type of pesticide that controls fungal disease by specifically inhibiting or killing the fungus causing the disease. Not all diseases caused by fungi can be adequately controlled by fungicides. These include the vascular diseases Fusarium and Verticillium wilt. Diseases caused by other types of organisms, disorders caused by abiotic factors, and insect damage are not controlled by fungicides. Thus it is essential to first determine the cause of symptoms before applying a fungicide.

Why are Fungicides Needed?

Diseases are a common occurrence on plants, often having a significant economic impact on yield and quality, thus managing diseases is an essential component of production for most crops. Broadly, there are three main reasons fungicides are used: (a) To control a disease during the establishment and development of a crop. (b) To increase productivity of a crop and to reduce blemishes. Diseased food crops may produce less because their leaves, which are needed for photosynthesis, are affected by the disease. Blemishes can affect the edible part of the crop or, in the case of ornamentals, their attractiveness, which both can affect the market value of the crop. (c) To improve the storage life and quality of harvested plants and produce. Some of the greatest disease losses occur post-harvest.

Fungi often spoil (render unusable) stored fruits, vegetables, tubers, and seeds. A few which infect grains produce toxins (mycotoxins) capable of causing severe illness or even death in humans and animals when consumed. Fungicides have been used to reduce mycotoxin contamination in wheat affected by Fusarium head blight, but most fungicides developed so far have not been sufficiently effective to be useful for managing mycotoxins associated with other diseases.

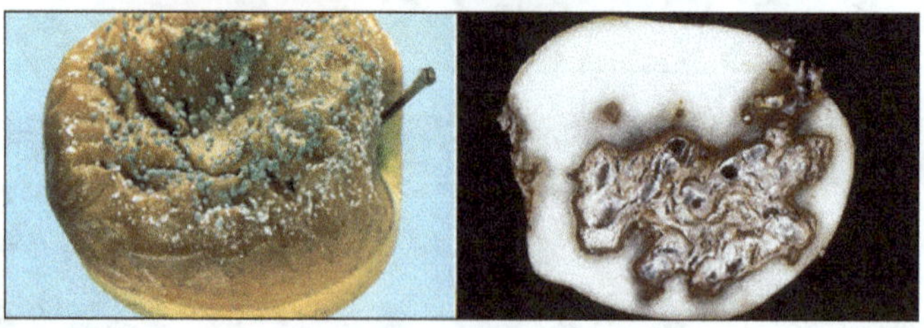

Role of Fungicides in Disease Management

Plant diseases are best managed by integrating a number of control practices that may include: crop rotation, selection of disease-tolerant or disease-resistant crop cultivars (cultivars genetically less susceptible than other cultivars), time of planting, level of fertilization, micro-climate modification, sanitation, and application of fungicides. Fungicides are often a vital part of disease management as (a) they control many diseases satisfactorily, (b) cultural practices often do not provide adequate disease control, (c) resistant cultivars are not available or not accepted in the marketplace for many diseases, and (d) certain high value crops have an extremely low tolerance for disease symptoms.

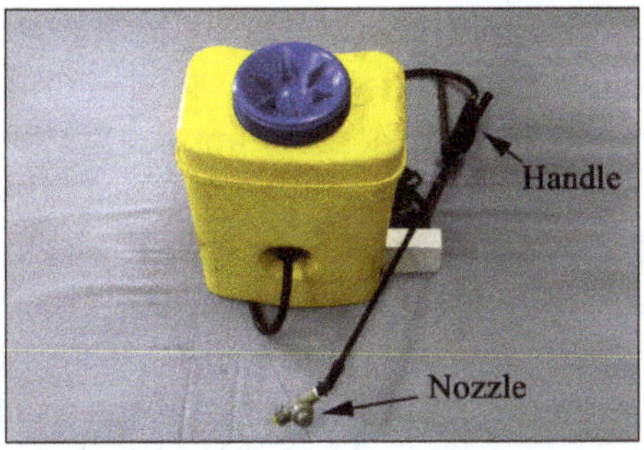

In contrast with most human medicines, most fungicides need to be applied before disease occurs or at the first appearance of symptoms to be effective. Unlike with many diseases of humans and animals, applying fungicides cannot heal symptoms already present, even if the pathogen is killed. This is because plants grow and develop differently than animals. Fungicides typically only protect new uninfected growth from disease. Few fungicides are effective against pathogens after they have infected a plant. Those that do have "curative" properties, which means they are active against pathogens that have already infected the plant, have limited ability to do so, often only being active on a pathogen within a few days of infection.

Many fungicides have targeted activity that imparts high efficacy against specific pathogens, which means low potential for toxicity to humans and other organisms, but also results in a high risk of pathogens developing resistance to the fungicide. A resistant pathogen is less sensitive to the action of the fungicide, which results in the fungicide being less effective or even ineffective. Fungicides that are designed to target specific enzymes or proteins made by fungi do not damage plant tissue, thus they can penetrate and move inside leaves enabling curative properties and increasing the amount of plant tissue protected to more than just where fungicide was deposit when applied. Since the mode of action of these fungicides is so specific, small genetic changes in fungi can overcome the effectiveness of these fungicides and pathogen populations can become resistant to future applications. Disease management strategies that rely heavily upon curative application of fungicides often lead to more resistance problems due to (a) the large size of the pathogen population when the application is made from which resistant individuals are being selected and (b) the difficultly in eradicating a pathogen entirely from inside the plant.

Growers often use disease forecasting systems or action thresholds, when these are available, to ensure fungicides are applied when needed and to avoid the expense and possible environmental impact of unnecessary applications. Forecasting systems have been developed for a number of diseases based on an understanding of the environmental conditions favorable for their development. Typically these are based on temperature and relative humidity or leaf wetness in the area where the crop is grown. Threshold-based fungicide programs involve routinely scouting the crop for symptoms, then applying fungicides when the amount of symptoms reaches a critical level beyond which the disease cannot be controlled adequately. An example of a critical level is one disease spot per five leaves examined. Knowledge of the disease cycle of the pathogen is important when developing and using forecasting systems and thresholds. Important aspects of the disease cycle include whether the disease is monocyclic (one generation per year) or polycyclic (multiple generations) and latent period (time between infection and production of new inoculum).

Economics often influence the choice of fungicide and application timing. Expensive fungicides and numerous applications are used on valuable plantings that might incur substantial economic loss in the absence of treatment, such as fruit trees and golf courses. Recognizing that with some diseases crop yield is not impacted when severity is low, an economic threshold is used to determine when fungicide treatment is needed. The crop tolerance level, or damage threshold, can vary depending upon the stage of the crop development when attacked, crop management practices, location and climatic conditions.

Application Methods

Fungicides are applied as dust, granules, gas, and, most commonly, liquid. They are applied to:

- Seed, bulbs, roots of transplants, and other propagative organs. These treatments are usually done by the seed company. Some treatments need to be done by the grower on-site at the time of planting. The goal is to kill pathogens that are on the planting material or to protect the young plant from pathogens in the soil.
- Soil either in-furrow at planting, after planting as a soil drench (including through drip irrigation), or as a directed spray around the base of the plant.
- Foliage and other aboveground parts of plants by means of a sprayer.
- Inside of trees via trunk injection.
- Harvested produce, as a dip or spray in the packinghouse.

Fungicides are used as a formulated product consisting of an active ingredient plus inert ingredients that improve the performance of the product. Fungicides are typically mixed with water then applied by spraying. Application equipment ranges from small hand-held and back-pack sprayers to large spray units carried by tractors or aircraft. A few fungicides are applied as dusts. Fungicides can also be applied in greenhouses as smoke, mist, fog or aerosol. Coverage of all parts of the plant susceptible to the disease is critical because very few fungicides can move adequately throughout a plant. Advancements are continually being made to nozzles and sprayers to improve coverage.

For many diseases, effective control necessitates multiple applications of fungicides, sometimes as frequently as every 5 days. Repeated applications are needed to protect new growth and to replace fungicide lost from the plant by chemical decomposition, UV-light degradation, and erosion by wind and water.

Types of Fungicides

Fungicides are categorized in several ways based on different characteristics:

Mobility in a Plant

Contact fungicides (also called protectants) remain on the surface of plants. Many contacts are potentially phytotoxic (toxic to plants) and can damage the plant if absorbed. Systemic (also called penetrants and mobile fungicides) are absorbed into plants. Most systemic move very short distances from the site of application, such as across a leaf blade from one surface to the other (local systemic or translaminar). Some fungicides are weakly systemic and can move further from the application site than local systemic, reaching all parts of the leaf the fungicide is deposited on.

A few systemics move more extensively because they are mobile in xylem tissue. When applied to the root zone, these are absorbed by roots and then move upward through the plant with the transpiration stream (xylem-mobile systemic). Xylem-mobile systemics applied to leaves move throughout the leaf where deposited, but cannot be redistributed out of that leaf; however, any material deposited on stems can move upwards into leaves. Phloem-mobile systemics (also known as "true" or amphimobile systemics) have bi-directional mobility, some material moves in phloem out of the leaf where deposited upwards to other leaves and downwards to roots. Systemics cannot move again after translocation.

Role in Protection

Contacts are products suited for preventive (prophylactic) use as they work by contact action on the surface of the plant to which they have been applied. Repeated applications are needed to protect new growth of the plant and to replace material that has been washed off by rain or irrigation, or degraded by environmental factors such as sunlight. Sometimes contacts are referred to as "residual" products because the deposited fungicide remains on the plant surface, occasionally as a visible residue, for several days. Due to their ability to penetrate plants, some systemics possess both preventive and curative (eradicant or kick-back) activity, thus affecting the pathogen after infection.

Breadth of Activity: Single-Site or Multi-Site

Single-site fungicides are active against only one point in one metabolic pathway in a pathogen or against a single critical enzyme or protein needed by the fungus. Since single-site fungicides are highly specific in their toxicity, having little effect on most organisms, they can safely be absorbed into plants, thus these fungicides tend to have systemic properties. As a result of this specific activity, fungi are more likely to become resistant to the fungicide because a single mutation in the pathogen usually allows it to overcome the action of the fungicide, such as by preventing it from binding to the active site in the fungus.

Typically, older contact fungicides have multi-site activity and thus usually affect many fungi in different classes. Through the development of in vivo screens, and due to the increase in the stringency and number of regulatory tests required to register a new active ingredient, fungicide manufacturers have found it easier to develop single-site systemics recently. As a result, fungicide resistance has become a more important concern in disease management.

Mode of Action

Fungicides kill fungi by damaging their cell membranes, inactivating critical enzymes or proteins, or by interfering with key processes such as energy production or respiration.

Others impact specific metabolic pathways such as the production of sterols or chitin. For example, phenylamide fungicides bind to and inhibit the function of RNA polymerase in oomycetes, while the benzimidazole fungicides inhibit the formation of beta tubulin polymers used by cells during nuclear division.

Some recently developed products are unique in that they do not directly affect the pathogen itself. Many of these elicit a response from the host plant known as "systemic acquired resistance" (SAR). These SAR inducers basically mimic chemical signals in plants that activate plant defense mechanisms such as the production of thicker cell walls and anti-fungal proteins. The utility of SAR inducers, however, has been limited so far since many pathogens are capable of over-powering such defenses.

Knowledge of exactly how a fungicide affects a fungus is helpful for selecting products. First, mode of action determines which fungi will be affected by a fungicide and thus which diseases can be controlled by using the fungicide. Second, fungicides with different modes of action are needed in a disease management program to delay fungicide resistance development.

Breadth of Activity: Narrow-Spectrum or Broad-Spectrum

Narrow-spectrum fungicides are effective against only a few usually closely related pathogens. These usually have single-site activity and are often systemic. Broad-spectrum fungicides can often control a wide range of unrelated pathogens. These usually are contacts with multi-site activity, but some have single-site activity. Several fungicides have activity that places them on a continuum between narrow and broad-spectrum fungicides.

Type of Chemicals

Fungicides also can be classified based upon their chemical composition. Chemically, organic molecules are those that contain carbon atoms in their structure whereas inorganic molecules do not. Many of the first fungicides developed were inorganic compounds based on sulfur or metal ions such as copper, tin, cadmium and mercury that are toxic to fungi. Copper and sulfur are still widely used. Most other fungicides used today are organic compounds and thus contain carbon. The term "organic" as used here is based on chemistry terminology and differs from "organic" used to describe a system of agriculture that strives to be holistic and to enhance agro-ecosystem health.

- Active ingredient (a.i.) is the active component of a pesticide, and is that moiety which is patented, synthesized, and registered by the manufacturer.
- Biofungicide is a naturally based microbial or biochemical product. There are three types of biopesticides. (a) Microbial biopesticides have an active ingredient that is a biological control agent (organism capable of attacking or competing with a pathogen or pest). (b) Plant biopesticides or plant-incorporated

protectants are defined by the EPA as "pesticidal substances that plants produce from genetic material that has been added to the plant". (c) Biochemical biopesticides contain naturally-occurring substances. Some biochemicals may also be products of fermentation. Biochemicals can directly affect disease-causing organisms or may stimulate SAR. Biopesticides generally are narrow-spectrum, have low toxicity, decompose quickly, and thus are considered to have low potential for negative impact on the environment. Many biofungicide products are being approved for organic crop production. While many have low toxicity, biopesticides are not necessarily safer than pesticides containing synthetic ingredients.

- Chemical group or class is the name given to a group of chemicals that share a common biochemical mode of action and may or may not have similar chemical structure.
- Chemical name is denoted by a nomenclature system designed by the International Union of Pure and Applied Chemistry (IUPAC); this set of rules is used to name organic, carbon-based compounds.
- Common name is the name proposed by the basic manufacturer for the active ingredient. This name is ratified by a committee either at IUPAC or the ACS (American Chemical Society).
- Reduced risk is a special classification approved by the EPA for specific uses of pesticides that have low risk to human health, low toxicity to non-target organisms, low potential to contaminate water or other environmental resources, and/or that broaden the adoption and effectiveness of integrated pest management strategies. Registration is expedited for uses designated as reduced risk.
- Formulation is the pesticide product. It is a prepared mix of active ingredient plus ingredients that improve pesticidal performance, such as carriers, solvents, wetting agents, spreaders, and stickers. Types of formulations include wettable powders, dry flowables, emulsifiable concentrates, and granules.
- Trade name/Trademark is the patented name under which a product is sold to the end user.

Fungicide Resistance

Fungicide resistance is a stable, heritable trait that results in a reduction in sensitivity to a fungicide by an individual fungus. This ability is obtained through evolutionary processes. Fungicides with single-site mode of action are at relatively high risk for resistance development compared to those with multi-side mode of action. Most fungicides being developed today have a single-site mode of action because this is associated with lower potential for negative impact on the environment, including non-target organisms.

When fungicide resistance results from modification of a single major gene, pathogen subpopulations are either sensitive or highly resistant to the pesticide. Resistance in this case is seen as complete loss of disease control that cannot be regained by using higher rates or more frequent fungicide applications. This type of resistance is commonly referred to as "qualitative resistance".

When fungicide resistance results from modification of several interacting genes, pathogen isolates exhibit a range in sensitivity to the fungicide depending on the number of gene changes. Variation in sensitivity within the population is continuous. Resistance in this case is seen as an erosion of disease control that can be regained by using higher rates or more frequent applications. Long-term selection for resistance in the pathogen by repeated applications may eventually result in the highest labeled rates and/or shortest application intervals not being able to adequately control the disease. This type of fungicide resistance is commonly referred to as "quantitative resistance".

Fungal isolates that are resistant to one fungicide are often also resistant to other closely-related fungicides, even when they have not been exposed to these other fungicides, because these fungicides all have similar mode of action. This is called cross resistance. Fungicides with the same Group Code are likely to exhibit cross resistance. Occasionally negative cross resistance occurs between unrelated fungicides because the genetic change that confers resistance to one fungicide makes the resistant isolate more sensitive to another fungicide.

Managing fungicide resistance is critically important to extend the period of time that an at-risk fungicide is effective. The primary goal of resistance management is to delay its development rather than to manage resistant fungal strains after they have been selected. Therefore, resistance management programs need to be implemented when at-risk fungicides first become available for commercial use. The objective of resistance management is to minimize use of the at-risk fungicide without sacrificing disease control. This is accomplished by using the at-risk fungicide with other fungicides and with non-chemical control measures, such as disease resistant cultivars, in an integrated disease management program.

It is critical to use an effective disease management program to delay the build-up of resistant strains. At-risk fungicides should be used at the manufacturer's recommended rate (full rate) and application interval. Using full rates is expected to minimize selection of strains with intermediate fungicide sensitivity when resistance involves several genes (quantitative resistance). At-risk fungicides should be used in alternation with other at-risk fungicides with different modes of action or different chemical groups, and they should be combined or alternated with fungicides that have a low resistance risk.

When one crop could serve as a source of inoculum for a subsequent crop, the alternation scheme among at-risk fungicides should be continued between successive crops such that the first at-risk fungicide applied to a crop belongs to a different cross-resistance group than the last at-risk fungicide applied to the previous crop. Some at-risk fungicides are formulated as premix products with other fungicides to manage resistance. At-risk fungicides should be used only when needed most. The most critical time to use them for resistance management is early in an epidemic when the pathogen population is small. Multi-site contact fungicides should be used alone late in the growing season, where they have been shown to provide sufficient disease control to protect yield. Another important component of resistance management is assessing disease control and reporting any loss of efficacy potentially due to resistance.

To promote resistance management, companies registering fungicides are voluntarily putting on the labels guidelines developed recently by EPA through a joint effort with the Canadian Pest Management Regulatory Agency (PMRA) under the North American Free Trade Agreement (NAFTA).

Legislation and Public Health

The two major laws governing fungicides and other pesticides in the United States are the Federal Insecticide, Fungicide, and Rodenticide Act (FIFRA), and Federal Food, Drug, and Cosmetic Act (FFDCA).

FIFRA was passed by Congress in 1947. Primary responsibility for its enforcement was initially with the United States Department of Agriculture (USDA), then transferred to the EPA in 1970. The Office of Pesticide Programs of EPA is chiefly responsible for regulating pesticides today. All pesticides must be registered before they can be sold and used in the U.S. unless their active and inert ingredients are deemed sufficient low risk to not require FIFRA regulation. To obtain registration, manufacturers of a pesticide must demonstrate that it does not have the potential to cause an adverse impact on the environment or non-target organisms, including people. This requires conducting several defined toxicology tests and investigating environmental fate. Additionally, the EPA must ensure that no endangered or threatened species or their habitat is harmed through use of registered pesticides. This assures compliance with the Endangered Species Act (ESA) of 1973 which prohibits any action that can adversely affect these species. In addition to federal registration with EPA, all pesticides must be registered with appropriate agencies in each state before they can be used.

FFDCA regulates the establishment of pesticide tolerances, which are the maximum permissible level of pesticide residues allowed in or on commodities for human food and animal feed. Manufacturers must include residue data in their registration materials. The Delaney Clause to FFDCA prohibited the presence in food of additives, including pesticides, considered carcinogenic. While well-intended, implementing this amendment became difficult as technology improvements enabled detection of additives at extremely low concentrations that were well below the dose necessary to cause cancer. Paradoxically, alternative pesticides could be allowed although they posed higher risks, if these were non-cancer risks.

The Food Quality Protection Act (FQPA) passed in 1996 replaced the Delaney Clause with a new health-based standard for evaluating food-use pesticides that includes a 'reasonable certainty of no harm' provision. Under the new standard, EPA establishes tolerances by considering (a) aggregate exposure to a pesticide from food as well as residential and other non-food uses, (b) cumulative effects to human health from other pesticides with a common mode of toxicity, (c) potential of increased sensitivity of infants and children as compared to adults, and (d) effect of the pesticide on estrogen and the endocrine system. EPA is reevaluating all existing pesticide tolerances under FQPA. As a consequence of FQPA and stricter EPA standards for pesticide registration, some older pesticides are not being re-registered and it is more difficult to register new products.

Integrated Pest Management

Integrated Pest Management (IPM) is an effective and environmentally sensitive approach to pest management that relies on a combination of common-sense practices. IPM programs use current, comprehensive information on the life cycles of pests and their interaction with the environment. This information, in combination with available pest control methods, is used to manage pest damage by the most economical means, and with the least possible hazard to people, property, and the environment.

The IPM approach can be applied to both agricultural and non-agricultural settings, such as the home, garden, and workplace. IPM takes advantage of all appropriate pest management options including, but not limited to, the judicious use of pesticides. In contrast, organic food production applies many of the same concepts as IPM but limits the use of pesticides to those that are produced from natural sources, as opposed to synthetic chemicals.

How do IPM Programs Work?

IPM is not a single pest control method but, rather, a series of pest management evaluations, decisions and controls. In practicing IPM, growers who are aware of the potential for pest infestation follow a four-tiered approach. The four steps include:

- Set Action Thresholds: Before taking any pest control action, IPM first sets an action threshold, a point at which pest populations or environmental conditions indicate that pest control action must be taken. Sighting a single pest does not always mean control is needed. The level at which pests will either become an economic threat is critical to guide future pest control decisions.
- Monitor and Identify Pests: Not all insects, weeds, and other living organisms require control. Many organisms are innocuous, and some are even beneficial.

IPM programs work to monitor for pests and identify them accurately, so that appropriate control decisions can be made in conjunction with action thresholds. This monitoring and identification removes the possibility that pesticides will be used when they are not really needed or that the wrong kind of pesticide will be used.

- Prevention: As a first line of pest control, IPM programs work to manage the crop, lawn, or indoor space to prevent pests from becoming a threat. In an agricultural crop, this may mean using cultural methods, such as rotating between different crops, selecting pest-resistant varieties, and planting pest-free rootstock. These control methods can be very effective and cost-efficient and present little to no risk to people or the environment.
- Control: Once monitoring, identification, and action thresholds indicate that pest control is required, and preventive methods are no longer effective or available, IPM programs then evaluate the proper control method both for effectiveness and risk. Effective, less risky pest controls are chosen first, including highly targeted chemicals, such as pheromones to disrupt pest mating, or mechanical control, such as trapping or weeding. If further monitoring, identifications and action thresholds indicate that less risky controls are not working, then additional pest control methods would be employed, such as targeted spraying of pesticides. Broadcast spraying of non-specific pesticides is a last resort.

Principles of IPM

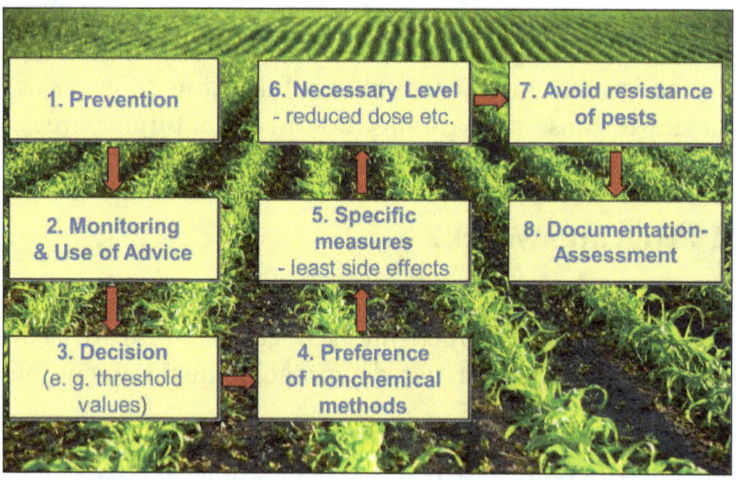

Principle 1: Prevention and Suppression

- Prevention is adoption of measures to reduce the chance of occurrence of pest. Suppression is reducing the impact of the pests.
- Prevention and suppression can be done by applying the different techniques.

- It is a method of preventing the spreading of harmful organisms by hygiene measures (e.g. by regular cleansing of machinery and equipment).
- One of the methods of prevention and suppression is crop rotation where it would break the life cycle of the pests.
- Prevention and suppression also include use of adequate cultivation techniques (e.g. stale seedbed technique, sowing dates and densities, under-sowing, conservation tillage, pruning and direct sowing).

Principle 2: Monitoring

- Harmful organisms must be monitored by adequate methods and tools, wherever available.
- Monitoring can be done through observations, use of scientifically sound warning, forecasting and early diagnosis systems, advice from professionally qualified advisers, etc.
- Many countries like France, Denmark have adopted this monitoring and forecasting technique.

Principle 3: Decision Making

- Decision making is done based on the results of the monitoring.
- IPM focuses on threshold-based intervention in most of the cases. Threshold is the defined pest density, or population level, which when exceeded, management should occur.
- However, threshold is difficult to define in most of the cases and in case of tolerant species, decision of intervention is based on the general observations.
- We should also be aware that specific crops, pest life cycle, climatic condition, etc., should be considered before making any kind of decisions.

Principle 4: Non-Chemical Methods

- Non-chemical methods are prioritized over chemical methods if they can produce satisfactory results.
- As chemical methods are often not sustainable and creates more pest problems, non-chemical methods are always preferred at first hand as they are more sustainable with less biological and environmental hazards.
- Examples of non-chemical methods include soil-solarization or biological control.

- Use of live natural enemies is one of the major non-chemical (biological) intervention method.
- Other non-chemical methods include biological, physical and ecological methods.

Principle 5: Pesticide Selection

- IPM doesn't totally avoid the use of the pesticides.
- When the alternative methods are not properly used then the pesticides are used for pest control.
- The pesticides used however need to be as specific as possible for the target.
- The pesticides shouldn't possess any threat to the health of human, non-target humans and environment.

Principle 6: Reduced Pesticide Use

- Reduced pesticides use refers to the reduction in the frequency and doses of the pesticides.
- This method needs to be supported by the other means of intervention.
- It helps in reducing the side effects of the pesticides.

Principle 7: Anti-Resistant Strategies

- IPM focuses on the anti-resistance activities as:
    - Unmanaged and haphazard use of the pesticides has created the problem of resistance.
    - Pests have developed the resistance and the uses of pesticides have less effect on them.
- This is also the major reason for the IPM.
- Anti-resistant strategies include use of combination of different pesticides that has different mode of action, applied in different time.

Principle 8: Evaluation

- Evaluation is the important aspect of the IPM program.
- Evaluation is done based on the records of the use of the pesticides, its effects and many more.

- Evaluation is necessary in studying the effectiveness of the plan protective measures and plan further.

Advantages of IPM

- Lower Cost Intervention:
    - Traditionally, the use of the pesticides to control the pest invasion would account to lots of cost.
    - Also, these pesticides need to be imported as well.
    - The application of IPM would lessen the financial burden.
    - Moreover, different techniques involved in IPM are more sustainable with long lasting benefits.
- Benefits to the Environment:
    - Use of the pesticides is often linked degradation of the environment causing some more additional problems.
    - IPM is an eco-friendly approach and the effect on the environment is always considered before the application of any interventions.
    - Less use of pesticides won't affect the fertility of soil.
- Minimizes Residue Hazards of Pesticides:
    - It is obvious that in an IPM schedule the use of pesticides will be considerably reduced, hence the pesticide residue hazards will also get automatically minimized.
- Anti-Resistance:
    - The IPM model in itself is the anti-resistant mode for pest control.
    - It discourages the use of chemicals and thus creates less cases of anti-resistance.
    - Pesticides are used only when the other alternatives are not satisfying.
- Useful and Best Intervention for the General Public:
    - Assurance of safe, reliable and low-cost pest control.
    - The pest control will not affect the crops.
    - Moreover, it is safe and affordable for the general public as well.

Disadvantages or Limitations of IPM

- More Involvement in the Technicalities of the Method:
    - IPM needs to be planned.
    - IPM demands more attention and dedication.
    - Requires expertise of various field.
    - All those involved in the IPM needs to be educated and trained which often requires much time.
- Time and Energy Consuming:
    - Application of IPM takes time.
    - Much time is needed in planning itself.
    - As IPM strategies differ from region to region, a separate plan is required for each region.
    - The expected results of intervention may take long time to be achieved.

Challenges in IPM

- Financial constraints in initiation of the IPM.
- Lack of involvement of the expertise.
- Inadequate research on pests.
- Lack of supportive government and policies.
- In developing countries where traditional farming is still practiced, farmers are unknown to the concept of IPM as well as IPM seems difficult to practice.

References

- Crop-agriculture: britannica.com, Retrieved 11, July 2020
- Crop-protection: byjus.com, Retrieved 29, March 2020
- Ware-intro-insecticides: ipmworld.umn.edu, Retrieved 12, January 2020
- Fungicides, disimpactmngmnt: apsnet.org, Retrieved 03, August 2020
- Integrated-pest-management-ipm-principles-advantages-and-limitations: publichealthnotes.com, Retrieved 17, May 2020

Chapter 2

Crop Diseases and their Management

Crop diseases refer to any alteration or harmful deviation in the normal functioning of the physiological processes of crops. This chapter discusses in detail the diseases in various crops such as diseases in lentils, diseases in chickpeas, diseases in field peas and diseases in wheat, etc.

Crop disease involves any harmful deviation or alteration from the normal functioning of the physiological processes. Therefore, diseased plants suffer disturbances from normal life processes and their vital functions.

Cabbage infected by bacterial blight disease.

In an attempt to reach high yields and healthy crops, farmers throughout the world struggle to prevent and eradicate various diseases from their crops. Each crop is susceptible to particular diseases that affect the quality and final yield potential.

Generally, it's estimated that various pests (insects, weeds, nematodes, animals, diseases) each year cause crop yield losses of 20-40%. More precisely, there is some data that maintains that crop diseases cause average yield losses of 42% for the most important food crops. In some cases, crop diseases destroy the whole crop production. For this reason, it's extremely important for farmers to find out all they can about the crop diseases so they can manage them properly.

Causes to Crop Diseases

The disease usually occurs and spreads from season to season, depending on the presence of a certain pathogen, as well as environmental conditions and the characteristics of each crop variety. Essentially, crop diseases occur according to the nature of their causal agent:

- Abiotic or Noninfectious disease agents.
- Biotic or Infectious disease agents.

Marginal leaves necrosis caused by freezing temperature.

Abiotic, or Noninfectious disease agents, include non-living environmental conditions or inappropriate farm management. They are not transmissible to other plants. A few universally recognized abiotic agents are:

- Extreme temperatures,
- Moisture,
- Wind,
- Frequent and heavy rain,
- Drought or flood,
- Excess or deficiency of nutrients,
- Soil compaction,
- Chemical injury caused by pesticides or salts,
- Improper water management.

Biotic, or Infectious disease agents, are living organism pathogens capable of spreading from one host to another and transmitting the disease.

Black or stem rust caused by the fungus Puccina graminis triciti.

The pathogens are classified as:

- Fungi, the most common pathogens, cause around 85% of plant diseases; examples include Black or stem rust caused by the fungus Puccina graminis tritici.
- Viruses are transmitted by a vector or attack the plant through a wound; for example, the Apple mosaic virus affects apple, plum, and hazelnut.
- Bacteria mutate and multiply rapidly; they enter the plant through a wound or stomata; for instance, the Apple fire blight is caused by Erwinia amylovora.
- Nematodes damage the crops causing galls on roots.
- Parasitic plants, live on crops, as they lack chlorophyll they obtain it from the host plant; for example, the dwarf mistletoe grows on other plants and derives nutrients from the host.
- Algae, theoretically don't cause significant damage, however, the may under special circumstances.

Healthy tomato root system (left) and one infected with nematodes (right).

Apple fire blight.

Apple mosaic virus.

How Disease Occurs in the Plant

Understanding the disease and the development process is the first step towards successful disease management. There are a few special conditions that are conducive for disease development. First, each crop is susceptible to some disease. Second, abiotic factors, namely weather, weaken the plant significantly. Therefore, the plant is vulnerable to the attack of the pathogen. So, the pathogen acts like a cherry on top. When all of the aforementioned factors (susceptible crop, abiotic stress, pathogen attack) are present together and at the same time, disease occurs. That is called the Plant Disease Pyramid.

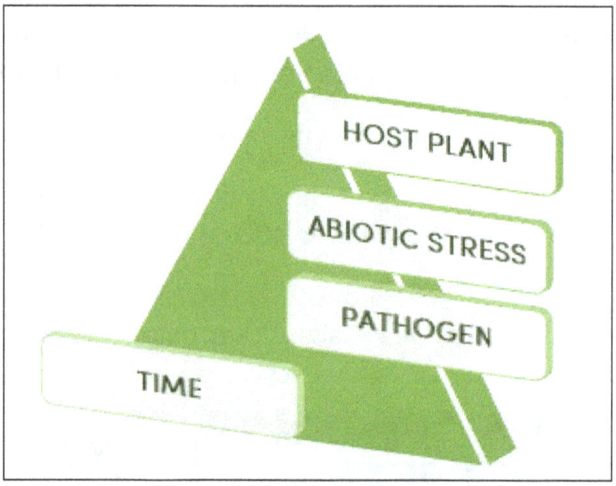

Plant disease pyramid.

Yields and Quality of Crops

Saving the yields and protecting the crops from diseases is a burning issue for every farmer who wishes to see the best results. From the beginning, the first and crucial step in crop protection is knowledge. Every farmer should be aware of susceptibility of the crop to certain diseases, as well as the abiotic agents that are favorable for the occurrence of the disease. More importantly, one of the most powerful practices is practicing preventative measures, such as:

- Planting resistant varieties.
- Managing optimal planting and harvesting times.
- Planting healthy and quality material.
- Disinfestation of equipment.
- Crop rotation.
- Managing plant nutrients according to crop needs.

Outside the preventative measures, there are also a few curative measures that should be taken in the case of the disease actually occurring. These measures include:

- The using of various chemicals, including fungicides, nematicides, bactericides, algacides.

- Removing infested plant parts.
- Utilizing various beneficial predators such as viruses, fungi, or bacteria.

There is one additional thing that could make a significant difference in disease farm management. Regular monitoring of crops and fields, as well as on-time reaction, can be a true lifesaver for every crop production.

Soil-Borne Diseases

In the case of soil-borne diseases, the pathogens can remain in the soil for long periods, waiting for the host - the plants - to come along. The environmental conditions can vary widely. Some pathogens favor damp conditions, some like certain soil pH levels and others target tender, succulent growth.

While some pathogens are short term visitors, appearing when the host and conditions are just right, others are naturally found in the soil and persist for years. When their favorite plant is not available, they may turn to an alternative. That's why clearing all the plant debris out of your garden at the end of the season is recommended.

Many plant diseases have similar symptoms, like yellowing leaves or dark spots. It is important to try and finds signs of the actual pathogen, but these are not usually visible without magnification. If you are having a longstanding problem, it would be worth your time to take a sample into your local Cooperative Extension.

Types of Soil-Borne Pathogens

Here are the common types of soil-borne pathogens:

- Fungi: The most common soil-borne pathogens. However, not every fungus causes plant problems and while the vast majority does not, over 8,000 fungi species do. And most plants are susceptible to some type of fungus.
    - Root rots cause the root system to begin to decay. The pathogens infect the plant's roots and block the uptake and flow of water and nutrients through the plant. Symptoms may include wilting, yellowing, stunting, dieback and eventual death and can be confused with other problems such as drought and nutrient deficiencies. Some common root rot fungi include Cylindrocladium, Pythium, Phytophthora, and Rhizoctonia.
    - Stem, collar, and crown rots affect the plant at ground level. Symptoms are similar to root rots, but since the rotting start above the soil line, it can be easier to detect early. Common pathogens to watch out for include: Phytophthora, Rhizoctonia, Sclerotinia, and Sclerotium.

- Wilt diseases, like Fusarium oxysporum and Verticillium spp. cause wilting of the plants, despite adequate water. There are also usually internal symptoms.
- Damping-off diseases affect young seedlings. They can be caused by a handful of fungi, including Pythium, Phytophthora, Rhizoctonia and Sclerotium rolfsii. They can infect the plants at germination or shortly after, causing sudden death. This is why the use of garden soil for seed starting is not recommended.

- Bacteria: Less common pathogens (and most don't stick around long). Some examples: Erwinia (soft rot), Rhizomonas (corky root of lettuce) Streptomyces (potato scab, soft rot of sweet potatoes).
- Viruses: Rare, thankfully, and most require living plant tissue to survive, but they can also hitch a ride on fungi or nematodes and flow in on water. When a virus enters a plant cell, it can cause the cell to produce more virus cells. Lettuce necrotic stunt virus affects Romaine lettuce plants, causing stunting and yellowing and sometimes spotting of lower leaves, while newer leaves remain green and thick.
- Nematodes: Sometimes called roundworms, nematodes are unsegmented worms with round bodies and points at both ends. Some are parasitic, like the nematodes sold to feed on beetle larvae in the lawn. And some will feed on or in roots. This is especially problematic for root crops, like carrots. Root rot nematodes are probably the most familiar. They cause distortion and swelling of roots and can affect the plant's vigor. Needle nematodes feed on the tips of roots, causing branching and swelling. And stubby root nematodes caused - yes - short, stubby roots.

White rot on garlic bulb, caused by soil-borne fungus.

Diseases in Lentils

Botrytis Grey Mold (Botrytis Cinerea and Botrytis Fabae)

Symptoms: Disease first appears as discrete cream colored lesions on lower leaves. Lesions girdle the stem and cover it with a furry layer of grey mold, eventually causing stem and whole plant death.

Factors favoring:

- High humidity and moderate temperatures with high moisture.
- Rain and sprinkler irrigation.
- Dense stands of lentils.

Botrytis injury.

Botrytis pod injury.

Anthracnose (Colletotrichum Truncatum)

Symptoms: Generally, symptoms appear later in the growing season. The disease kills lower leaves, causing defoliation and eventually premature plant death. Brown lesions with dark borders may be present on leaves and stems, sometimes with small black bristle-like fruiting bodies (pycnidia). Stem lesions could potentially grow to girdle the plants. Typical field symptoms include lodged plants with abnormal dark brown stems and pycnidia on the lower stem.

Factors favoring:

- Infected fields nearby.
- Humid to moist weather conditions, rain splash, wind, and sprinkler irrigation.
- Leaf wetness for 18 to 24 hours.
- Moderate temperatures ranging between 20 to 25 °C (68 to 75 °C).Dense stands of lentils.

Early infection.

Heavy Anthracnose infection showing black stems.

Fusariums Wilt (Fusarium Oxysporum f.sp. Lentis)

Symptoms: Disease appears in patches at seedling and adult stages. Seedling wilt is shown by sudden drooping, followed by drying of leaves and the whole seedling. Symptoms are seen from flowering to pod filling stages. Stunting of the plant, wilting of top leaves, yellowing of foliage and death of the plant are the obvious symptoms on the infected plants. A slight reduction of lateral root growth is seen in a healthy root system.

Factors favoring:

- Soilborne disease.
- Warm spring and hot, dry summer.
- Optimum temperature is around 25 °C (75 °F).

White Mold (Sclerotinia Sclerotiorum)

Symptoms: Leaves and stem turn brown to tan and die prematurely. Infected areas are covered with white fluffy mold. Dark brown to black scletotia develop inside and outside of the infected plants.

Factors favoring:

- Wet and cool weather conditions.
- Dense stands of lentils.

Ascochyta Blight

Causal Organism: Ascochyta lentis.

Ascochyta lesions on leaflets.

Symptoms: Spot-like lesions can develop on leaves, stems, pods, and seeds. Small black pycnidia become visible with maturity. Premature leaflet drop, stem breakage, and stem dieback may occur. Affected crops under severe infection may be blighted and seed may become shriveled, reduced in size, and discolored. Flowers and pods could abort, leading to yield loss.

Factors favoring: Non-resistant cultivars grown and planting low quality seed will increase the chances of infection. This pathogen does best in cool, wet weather. If crop rotation is not used, the severity of infection may increase. Residue from previous crops provides high inoculum risk. Infection may be increased if protected seed treatments and fungicides are not used. Also, not taking advantage of helpful cultural controls such as deep plowing, harvesting early, desiccating crops before harvest, and growing a non-host barrier grown in-between lentil crops and residues are a sure way to increase inoculum carryover.

Affected crops may be severely blighted and seed becomes shriveled and discolored.

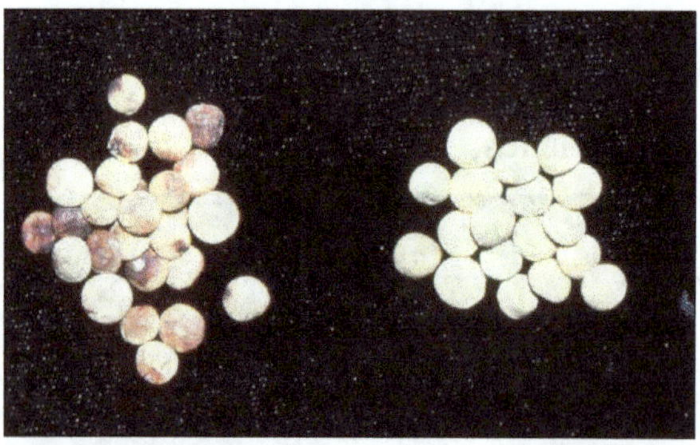

Ascochyta infected (on left) vs. clean seed (on right).

Root Rot Seedling Blight

Root rots and seedling blights in lentil are primarily caused by soil-borne fungi including Rhizoctonia, Fusarium, and Pythium species.

Rhizoctonia Sp.

Symptoms: Reddish to dark brown root lesions can develop on epicotyls and hypocotyls.

Brown discoloration occurring near the soil line on the epicotyl could girdle the stem. Seeds may rot, or when growing, develop rot or pre and post-emergence damping-off. Lesions developing on roots that eventually progress enough to pinch them off can be observed. Stem lesions can develop at or below the soil line that may expand above the soil line to the lower branches in older plants. Leaves can turn yellow from the base of the plant up. Circular patches of stunted plants may be observed in the field. A plant that can be easily pulled out from the soil is indicative of rot.

Factors favoring: No-till or reduced-till and sandy soil favors infection, particularly when organic matter is high. Cool temperatures (11-18 °C) and root knot nematode presence make plants more susceptible to infection, although the pathogen tolerates a wide range of soil temperatures. Infection is more likely when paired with diseased seed, compacted soil, herbicide carryover, and non-rotation of crops.

Rhizoctonia injured seedlings.

Fusarium Sp.

Symptoms: Patches are found in the field, usually occurring during flowering and on to the fruiting stage; however, infection may occur at any stage of the plant cycle. Plants may be stunted overall and the leaves often curl from the top to the bottom of the plant. Top leaves in the plant may wilt and droop. They can also shrink and curl without defoliating early. Some other symptoms include yellowing, reduced root system with discoloration, poorly developed nodules, and damage at the taproot tip. Seeds may rot, pre-emergence damping off can occur, and the plant may die.

Factors favoring: The pathogen is favored by high soil temperature and dry conditions. Presence of nematodes can increase severity of the infection. Also, using non-resistant seed or infected seed and having low levels of saprophytic microflora in the soil can increase infection.

Pythium Sp.

Symptoms: Plants may show stunting, yellowing or purpling of the leaves, and discoloration of the main tap root and feeder roots. Seeds may rot, damping-off can occur, and the lateral roots may develop poorly. If the environment becomes warm and dry, the plants may die. From a distance, the stands look unfavorable and the plants may turn beige to light brown in color and eventually die.

Factors favoring: Planting early in wet, cool soils may increase infection along with using low quality seed or nonresistant plants, infested debris from previous seasons, and planting on clay soils.

Stem Rot of Lentil (White Mold)

Sclerotinia Sclerotiorum

Symptoms: A plant infected with Sclerotinia can develop bleached lesions on stems, leaves, pedicels and pods. The tissues that have stem lesions may turn yellow and wilt. Patches of dead plants may be observed. The plant parts are often covered in white, cottony mold. Dark brown to black sclerotia may be found on the plant or on the soil under its canopy.

Factors favoring: Conditions that favor the disease include humid environment, cool conditions, and dense canopies. Infested seed and susceptible host debris from previous seasons increase infection.

Crop destruction by Sclerotinia.

Anthracnose

Colletotricum Truncatum

Symptoms: The disease usually appears in the lower leaves first as necrotic and tan to creamy brown lesions, leading to leaf drop occurring in severe cases. Stems can develop tan to brown lesions which may have a defined black border and create indents along the stem that can even coalesce and girdle the stems, resulting in wilting. From a distance, brown patches and lodged lentil plants are formed in the crop resulting in lower seed yields. Pods may turn brown and necrotic. Seeds that are affected may have brown discolored spots and become shriveled, although not all infected seeds show symptoms.

Factors favoring: Conditions conducive for infection include warm, humid, and wet environments, especially late in the season. Having a dense canopy, infested seed, and leaving crop debris infested from previous years are all ways to increase inoculum. Moderate temperatures of 20-24 °C (68 to 75 °F) are optimal. Tillage of anthracnose infested fields increase survival of the pathogen in areas with cold winters.

Early infection.

Heavy Anthracnose infection showing black stems.

Diseases in Field Peas

Downy Mildew

Disease Symptoms:

- A grayish white, moldy growth appears on the lower leaf surface, and a yellowish area appears on the opposite side of the leaf.

- Infected leaves can turn yellow and die if weather is cool and damp.
- Stems may be distorted and stunted.
- Brown blotches appear on pods, and mold may grow inside pods.

Favourable Conditions:

- High humidity and low temperatures (5- 15 °C) for few days are ideal for infection and development of disease.

Survival and Spread:

- Primary infection by soil, seed and water Secondary infection by sporangia through rain splash or wind.

Powdery Mildew

Disease Symptoms:

- It attacks leaves first producing faint, slightly discolored specks from which grayish white powdery growth of mycelium develop.
- Powdery growth spread over leaf, stem and pod.
- The leaves turn yellow and die.
- The fruits do not either set or remain very small.
- It causes defoliation.
- Later stages, powdery growth also covers the pods.

Survival and Spread:

- Powdery mildew spores are carried by air and once active, will continue to spread in dry conditions.

Favourable Conditions:

- Warm (temperature 15-25 °C), humid (over 70% relative humidity) conditions for 4-5 days late in the growing season, during flowering and pod filling, favour disease development.

Pea Rust

Disease Symptoms:

- Leaves of infected plants exhibit many small, orange-brown pustules usually at the lower surface.
- Severely infected leaves wither and may drop from the plant.
- Larger pustules occur on the stems and isolated pustules may be found on the pods.
- Severe infection may result in reduced seed size and may cause yield losses of up to 30%.

Survival and Spread:

- Euphorbia and infected vegetation residues are sources of the infection.
- The agent is not transferred by seeds.

Favourable Conditions:

- Frequent precipitations, plentiful dews and air temperature of 20-25 °C promote development.
- Dry and hot weather restrains the disease development.

Ascochyta Blight

Disease Symptoms:

- Early symptoms are most commonly observed under the plant canopy, on lower leaves, stems, and tendrils, where conditions are more humid.
- Symptoms first appear as small, purplish-brown and irregular flecks.
- Under continued humid conditions, the flecks enlarge and coalesce, resulting the lower leaves becoming completely blighted.
- Severe infections may lead to girdling of the stem near the soil line, which is known as foot rot.
- Foot rot lesions are purplish-black in colour and may extend above and below the soil line.
- Foot and stem lesions girdle and weaken the stem, leading to crop lodging and yield loss.
- Disease lesions develop on pods under prolonged moist conditions or if the crop has lodged.
- Pod lesions are initially small and dark, but may become extensive and lead to early pod senescence.
- Severe pod infection may result in small, shrunken or discoloured seed; or alternatively, seed may show no symptoms.

Survival and Spread:

- Ascospores carried long distances by wind.
- The asexual conidia travel short distances to new hosts via water splashes from rain.
- Infection originates from diseased seed or from spores growing on debris in the soil near pea plants.

Favourable Conditions:

- Favorable conditions are warm humid conditions with a temperature is about 15 °C to 25 °C.

White Rot

Disease Symptoms:

- The infection may occur at any part of the foliage, mainly the stem or branches.
- The maximum infection develops at the flowering stage of the crop, when petals fall on ground and these catch infection immediately and mycelial growth of fungus invades the stem and branches.
- At the point of infection, a dry discolored spot develops.
- It gradually girdles the entire stem and also progresses up and down.
- As a result of tissue necrosis, the portion of the plant beyond the point of infection wilts.
- If the infection is at the base of the main stem, the entire plant wilts.
- If it occurs on branches, partial wilting occurs.
- The diseased tissues become whitish and may be shredded.

Survival and Spread:

- The infection caused by hard sclerotia left in soil or in plant debris.

Favourable Conditions:

- When adequate moisture is available and temperature ranges between 4 and 20 °C. However, light is essential for stimulation of apothecial production.

Root Rot

Damage Symptoms:

- Reddish brown to black streaks appear on primary and secondary roots.

- These streaks coalesce at later stages, leading to girdling of lower stem.
- Red discoloration of the vascular system can be seen, especially near cotyledon attachment.
- Stunted growth, yellowing and necrosis appear on the basal foliage.

Survival and Spread:

- Primary infection by soil, seed and water.
- Secondary infection by conidia through rain splash or wind.

Favourable Conditions:

- Cool, wet weather conditions.
- Favourable conditions are higher soil temperatures 25° to 30 °C and moderate soil moisture.

Wilt

Disease Symptoms:

- Symptom of the disease is more pronounced in 3 to 5 week old plants. In young seedlings, cotyledons droop and wither.
- Yellowing of lower leaves and stunting of plants.
- The xylem vessels develop brown discoloration and get distorted.
- Leaflet margins curl downward and inward.
- The stem may be slightly swollen and brittle near the soil.
- Internal woody stem tissue often is discolored, turning lemon brown to orange brown.
- Externally, the root system appears healthy; however, secondary root rots are likely to occur on plants wilted for long periods.

- Eventually, wilted plants may die.

Survival and Spread:

- Primary infection by soil, seed, water.
- Secondary infection by Conidia through rain splash.

Favourable Conditions:

- A soil temperature of 23° to 27 °C is most favourable for Fusarium wilt.
- Hot weather and warm soils.

Root Knot Nematodes

Damage Symptoms:

- Root-knot nematode feed on tender roots and base of pseudostem causing stunting, chlorosis, poor tillering and necrosis of leaves are the common aerial symptoms.
- Characteristic root galls and lesions that lead to rotting are generally seen in roots.
- The infested roots have brown, water soaked areas in the outer tissues.
- Nematode infestation aggravates root rot disease.

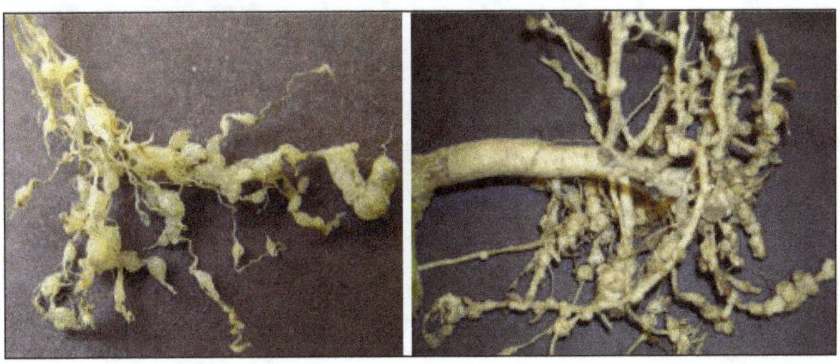

Survival and Spread:

- Nematodes survive in soil and infected roots as primary inoculums.
- Therefore, tissues from infected crops remaining in the field serve as a reservoir of the fungus.
- It spreads from infected plants or through soil.

Favourable Conditions:

- Warm, moist soil is favourable conditions.

Life Cycles of Diseases

- Downy Mildew

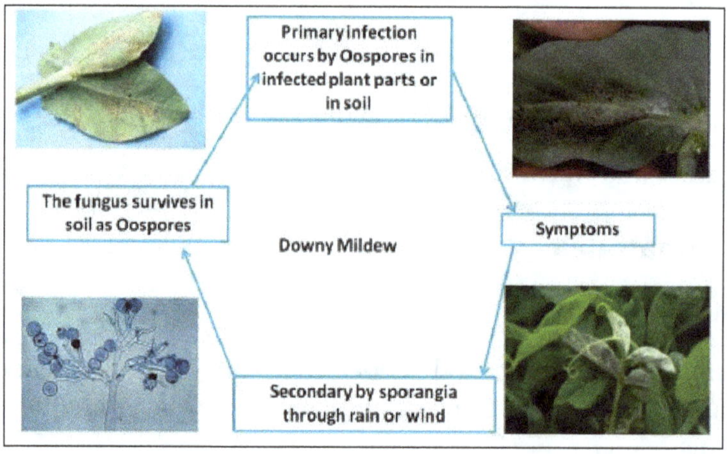

- Powdery Mildew

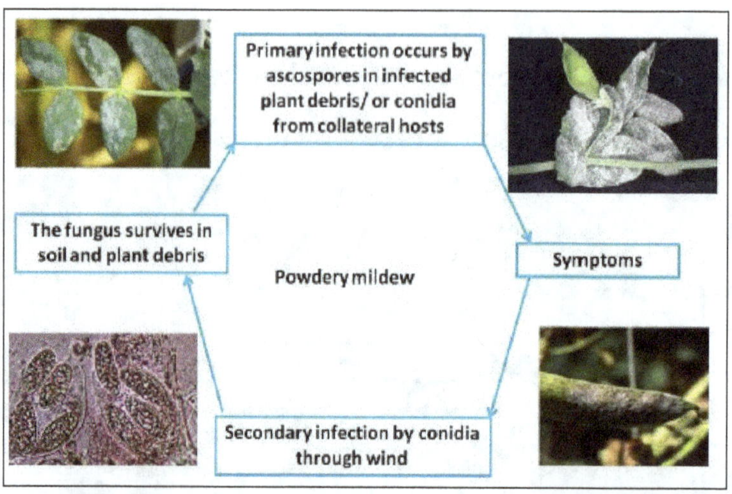

- Pea Rust

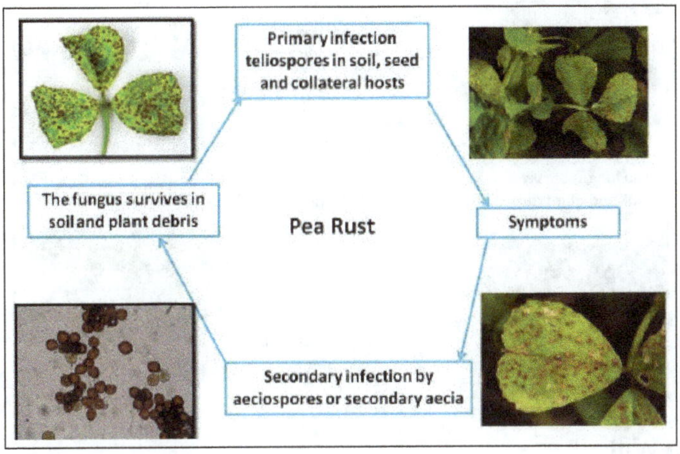

- Root Rot

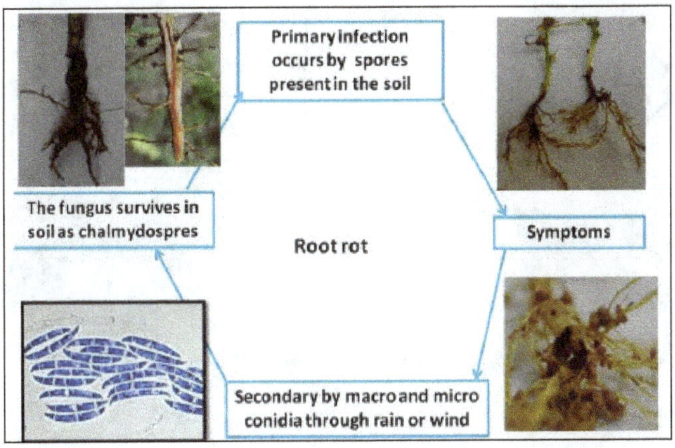

- Ascochyta Blight

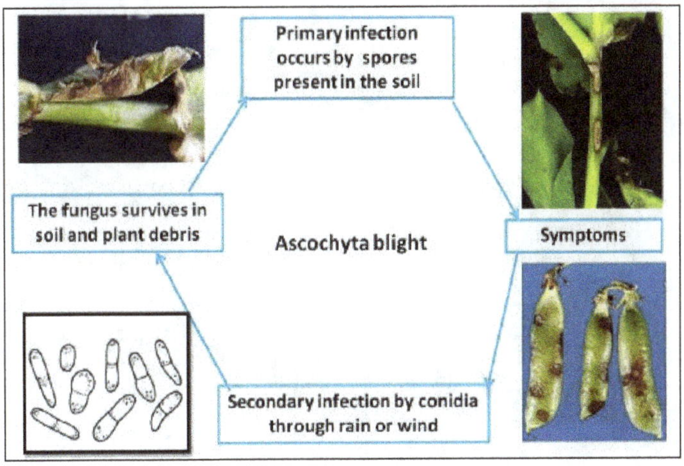

- White Rot

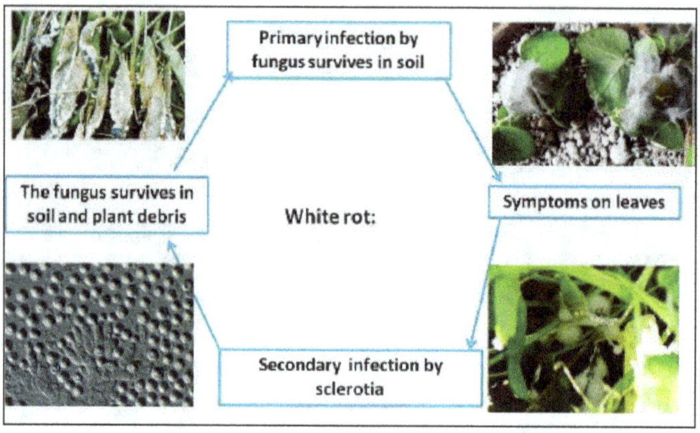

- Fusarium Wilt

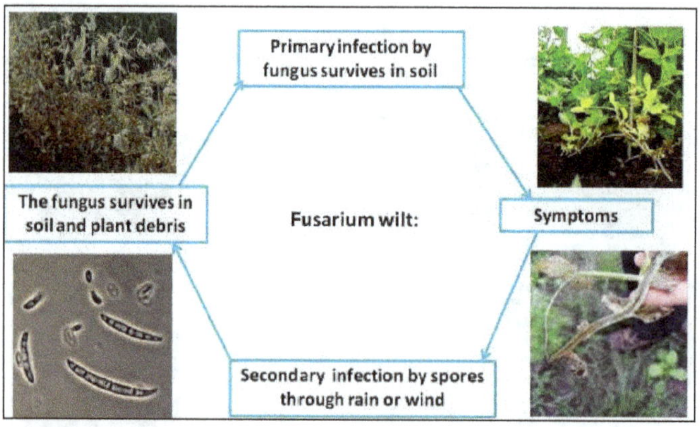

- Root knot Nematode

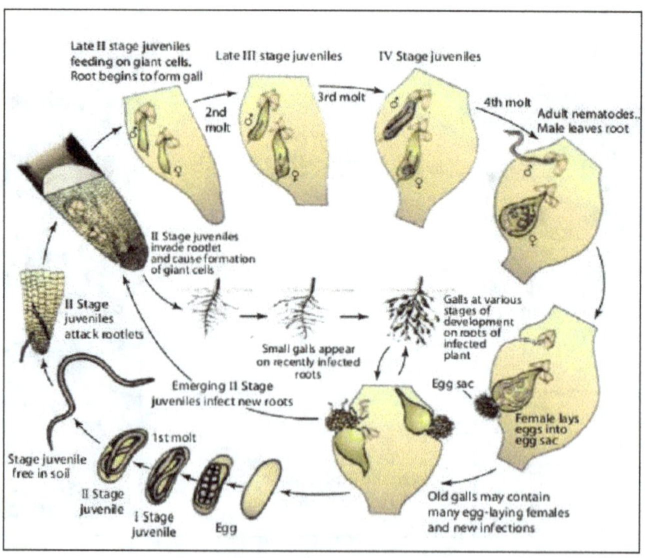

Rhizoctonia root disease.

Diseases in Faba Bean

Ascochyta Blight, Leaf Blight and Pod Spot

Scientific names: Ascochyta fabae Gossen, Sheard, C.J. Beauch. & Morrall [anamorph], (1986): Didymella fabaeG.J. Jellis & Punith, (1991) [teleomorph].

Significance

Yield losses of 10-30% can occur in seasons favorable for the disease. Discoloration of seed can seriously reduce its market value.

Symptoms

Symptoms occur on leaves, stems and pods of infected plants and can be confused with the early stages of chocolate spot. On leaves, small, circular, dark-brown spots appear first. As the disease develops, they enlarge and turn light and then change to dark grey in color. They become irregular in shape, often zonate, and may coalesce to cover most of the leaf surface. Leaf tissue next to the lesions may become black and necrotic. Within the lesions, numerous pinhead- sized black fruiting bodies (pycnidia) of the fungus develop. These appear only under moist conditions and are often concentrically arranged on the stem; lesions are more elongated, sunken and darker than leaf lesions and are usually covered with scattered pycnidia. Stems may split and break at the point of infection causing plants to lodge. On pods, lesions are sunken and have pale centre and dark margins; they can be covered by numerous pycnidia. Well-developed lesions can penetrate the pod and infect developing seeds causing them to be shrunken and discolored. Badly infected seeds have yellowish brown stains on the outer seed coat, which considerably reduces its market value.

Ascochyta blight can cause seed staining in pods close to maturity even when disease levels in the crop have been too low to warrant fungicide sprays. Faba bean seed that has greater than 25% seed coat discoloration can reduce the emergence of seed by 30%. Seed that has less than 5% seed coat discoloration will usually have normal levels of germination.

Hosts

Faba bean.

Geographic Distribution

Africa (Egypt, Morocco); America (Argentina, Canada (NS); Asia (China, Syria, Israel, Japan, Korea, Turkey); Australasia (Australia, New Zealand); Europe (Britain, Czechoslovakia, Denmark, Germany, Italy, Jersey, Norway,Russia).

Biology and Transmission

This fungal disease has an asexual and sexual stage; the asexual stage is most common. In this stage the fungus survives mainly on infected seed and on crop residues. Spores of the fungus produced on crop residues can be carried onto new crops by wind. Infection can occur at any stage of plant growth, provided conditions are favorable. Moisture is essential for infection to occur. During wet weather, the disease can spread further than in dry conditions because spores of the fungus are carried onto neighboring plants by wind and rain splash.

Detection/Indexing Method in Place at the CGIAR Center at ICARDA

- Malt Extract Agar.

Treatment/Control

Crop Rotation

- Since only Faba bean is susceptible to ascochyta blight caused by A. fabae inclusion of non-host crops in the rotation will bring about a reduction in the inoculum level.
- Faba beans should not be grown more frequently than every 3-4 years.
- Furthermore, new crops should not be grown near fields that were infested with blight in the previous year due to the risk of infection by air borne ascospores.

Tillage

Under most conditions, deep ploughing will hasten decomposition of infected Faba bean straw and remove it as a source of inoculum. Studies conducted in the Pallouse Region in Washington State showed that the pathogen survived in naturally infected plant material for more than 2 years when situated on the soil surface, but lost its viability rapidly at soil depths of 10-40 cm. All infested crop residues and volunteer plants should therefore be destroyed by thorough tillage and proper crop rotation.

Fungicide Application

Growers should take special care to protect green and healthy foliage during pod filling, especially if blight symptoms appear and increase in severity. Tests with application of Bravo 500 at different crop stages have shown that spraying at early podding reduced pod damage and markedly increased seed yields in the moderately resistant cultivars. Foliar fungicide applications may not prove cost effective when disease pressure is low. Thiram-based fungicides are registered for the treatment of grey mould infected seed.

While seed treatment can improve establishment, it does not provide any protection from air-borne infection later in the crop.

Procedure followed at the Centers in Case of Positive Test

Destroying seed for intensive spots, and seed treatment for no visible spots.

Ascochyta blight.

Chocolate Spot

- Scientific name: Botrytis fabae Sardina.
- Other scientific name: Botryotinia fabaeWu & Lu.

Significance

Loss depends on the severity of infection, the time at which infection occurred, and the amount of spring rainfall. In unprotected crops, the disease can reduce yields by at least 30-50% under conditions favorable for disease development. In addition, seed from badly affected plants may have a reddish-brown discoloration, which lowers its market value.

Symptoms

These are varied, and range from small spots on the leaves to complete blackening of the entire plant. Leaves are the main part of the plant affected, but under favorable conditions for the disease it also spreads to stems, flowers and pods. Two stages of the disease are usually recognized. First, a non-aggressive phase, when discrete reddish-brown spots are "peppered" over the leaves and stems, and then an aggressive phase, when spots darken in color and coalesce to form larger grey-brown target spots that may eventually cover the entire plant. Small black sclerotia may sometimes be found inside the stems of badly diseased plants.

Red-legged earth mite (RLEM) damage can be mistaken for chocolate spot. This starts as silvery patches which become red-brown, similar in color to chocolate spot but form large irregularly-shaped areas. Red-legged earth mite damage usually occurs during the seedling stage and on the lower leaves.

Factors such as phosphorus or potassium deficiency, water logging and excessive weed burdens which reduce crop vigor may make plants more susceptible to the development of chocolate spot. Leaves which have been damaged by insect attack or in wheel tracks are also more susceptible.

Hosts

Vicia spp. especially Vicia faba (field or broad bean); also reported from Phaseolus beans, pea and lentil.

Geographic Distribution

Europe, Middle East, N.E. and S. Africa, S.E. Asia, Australia, S. America. More recent records are from Korea, India, S. Australia, Canada and Norway.

Biology and Transmission

Chocolate spot, caused by Botrytis fabae and Botrytis cinerea can survive either as sclerotia in the soil or on crop debris, in infected seed, or on self-sown volunteer plants. In hitherto unaffected areas the disease often becomes established by the sowing of infected seed. In subsequent years, the initial infection usually occurs when spores formed on old bean trash are carried by wind into new crops. These spores may move long distances. Once the disease becomes established it rapidly spreads within a crop, and within 4-5 days of infection spores can be formed on infected tissue and initiate secondary spread of the disease. The fungus is most aggressive under cool, humid conditions, particularly at flowering time. Aggressive development of stem infection late in the season can cause the crop to lodge.

Air-borne conidia are readily produced on dead bean foliage during wet weather. The fungus perennates in bean debris and as sclerotia in the soil and is also seed-borne.

Detection/Indexing Method in Place at the CGIAR Center at ICARDA

- Agra Plate Test.

Treatment/Control

- Use of clean seed is advocated. Seed treatment with fungicides such as benomyl (0.2-0.1%) chlorothalonil (0.2-0.3%) or thiabendazol (0.1-0.2%) is recommended to control the seed- borne infection. Practice a 2-3 year crop rotation along with deep plowing to reduce soil-borne inoculum.
- Foliar spray using chlorothalonil (2-3 L\ha) as single application at early bloom to early pod set provides the best protection and increases seed yield.

Procedure followed at the Centers in Case of Positive Test

- Destroying serious infected seed and seed treatment for light infection.

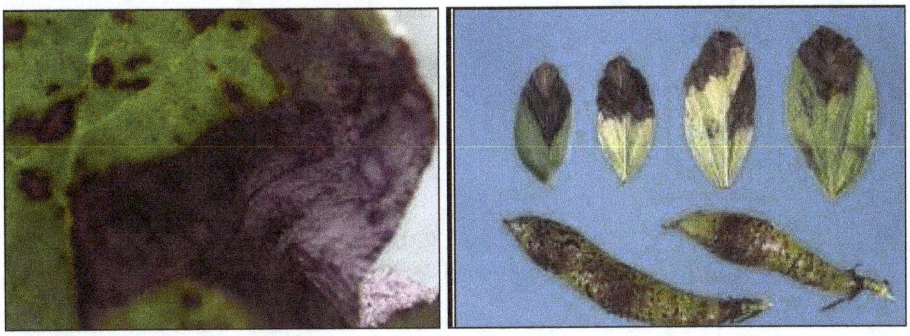

Chocolate spot.

Botrytis Grey Mold of Faba Bean

- Scientific name: Botrytis cinerea Pers. 1794.
- Other scientific name: Botryotinia fuckeliana (de Bary) Whetzel.

Significance

The disease is capable of causing serious yield losses in years when spring rainfall is high and/or there are prolonged wet periods especially if it is founded with B. fabae.

Symptoms

All aboveground plant parts of Faba bean can be affected by botrytis grey mould. Depending on the location of the crop, symptoms may initially appear either on flowers and pods, or lower in the crop canopy. The most damaging symptoms become apparent after the crop has reached canopy closure and a humid microclimate is produced under the crop canopy. The disease appears first as discrete cream colored lesions on lower leaves. These enlarge and coalesce to infect whole leaflets which later senesce and fall to the ground. Unlike ascochyta blight, no small, black fruiting bodies (called pycnidia) can be seen within the lesions. If conditions remain conducive for disease, that is warm and wet under the crop canopy for at least 4 days, infection can spread to the lower stems. These lesions will girdle the stem and become covered with a furry layer of grey mould, eventually causing stem death and whole plant death, often occur before the onset of flowering and pod fill. Infection will continue to spread resulting in patches of dead plants within crops. Pods which become infected will be covered in a grey moldy growth, rot, and turn brown when dried out.

Hosts

Botrytis cinerea has a broad crop host range collectively, including faba bean, chickpea, field pea, lupin and pasture legumes such as lucerne and clover. Other host species include a wide range of ornamental and horticultural crops.

Geographic Distribution

Cosmopolitan.

Biology and Transmission

The fungal pathogens Botrytis cinerea that cause botrytis grey mould can survive as several forms, these include in infected seed, sclerotia in the soil, in old infected trash, and on alternate host plants. Sowing seed that is infected by the botrytis grey mould pathogens can give rise to infected seedlings and the appearance of seedling blight symptoms, which can reduce seedling survival and reduce crop establishment. Old infected trash is an important source of fungal inoculum. Spores are produced on old trash and are carried by the wind into new crops where infection can occur.

Detection/Indexing Method in Place at the CGIAR Center at ICARDA

- Agar Plate Test.

Treatment/Control

- Use of clean seed is advocated. Seed treatment with fungicides such as benomyl (0.2-0.1%) chlorothalonil (0.2-0.3%) or thiabendazol (0.1-0.2%) is recommended to control the seed- borne infection. Practice a 2-3 year crop rotation along with deep plowing to reduce soil-borne inoculum.

- Foliar spray using chlorothalonil (2-3 L\ha) as single application at early bloom to early pod set provides the best protection and increases seed yield.

Procedure in Case of Positive Test

- Seed treatment.

Anthracnose

- Scientific name: Colletotrichum lindemuthianum (Sacc. & Magnus) Briosi & Cavara.

- Other scientific name: Colletotrichum truncatum (Schwein.) Andrus & Moore.

Significance

Not Significant.

Symptoms

- Leaf lesions and premature leaf drop: In most faba bean crops, the first symptoms of anthracnose appear before flowering, when the plants have 8 to 12 nodes on the main stem. This is also the time when the first tendrils form, and approximately a week before flowers start to open. If there is a large amount of inoculum in the field the first symptoms may appear earlier. Tan coloured lesions of variable size develop on the lower leaflets and the most severely affected leaflets die and drop to the ground. This premature leaf drop indicates that anthracnose may become a problem, and fungicide application should therefore be considered.
- Stem lesions: Lesions on stems develop soon after the appearance of leaf lesions, primarily at the base of the plant. Stem lesions may be small, brownish with a black border, or larger, stretching along the stem. As the season progresses, more and more lesions develop at the stem base, as well as on the upper part of the stems, and many stems are girdled.
- Wilt: Anthracnose causes defoliation and stem girdling, which inhibits utilization of water and nutrients, and causes the faba bean plants to wilt. As a result, large areas of brown and dying plants can be found in the field.

Hosts

The fungus Colletotrichum truncatum causes anthracnose in lentil and only attacks plant species belonging to the Lens and Vicia family, such as fababean and wild vetches. This means that a few weedy species in Canada can harbor the disease, but other crops, except fababean, are not at risk.

Geographic Distribution

It is reported from Bangladish, Canada, Ethiopia, Morocco, and Syria. It is economically important only in Canada.

Biology and Transmission

Small, pinhead sized fungal structures (microsclerotia) form on the infected plant tissue. They may be seen with the unaided eye in the centre of stem lesions or more easily with a hand lens (10-15 x magnification). Each microsclerotia consists of a few hundred cells with thick, black cell walls that protect the fungus from colonization by other microorganisms. Microsclerotia enable the fungus to survive between faba bean crops either on the plant debris or free in the soil. They remain viable longer when buried in the soil by tillage than left exposed to weather extremes on the soil surface.

Detection/Indexing Method at ICARDA

- Agar Plate Test.

Treatment/Control

- Use either certified seed, approved seed, or seed known to have a long disease-free history. The use of disease-free seed is the most important control measure.
- Do not plant beans for at least two years in land that has carried an infected crop.
- Remove diseased plants, where practical, to help check the spread of disease.
- Avoid cultivating and harvesting an affected crop when wet to prevent the spread of spores.
- Do not pack lightly diseased pods as anthracnose can develop during transport.
- Bravo 500 (50% cholorothalonil, Zeneca Agro). The recommended rate is 0.8 - 1.6 L/acre (2.0-4.0 L/hectare) with a maximum of two applications in a season. The water rate is 90-640 L/acre (220-1600 L/hectare). Bravo must be present on the plant surface prior to the onset of fungal infection. It sticks well to the plant surface and resists removal by rain. A second application 10-14 days later may be necessary under wet weather conditions and to protect new growth. Fungal resistance to cholorothalonil has not been detected.

Procedure followed in Case of Positive Test

- Seed treatment.

Anthracnose

Phoma Blight, Spring Black Stem and Leaf Spot

- Scientific name: Phoma medicaginis var. medicaginis Malbr. & Roum, 1886.
- Other scientific name: Phoma medicaginis.

Significance

Not Significant.

Symptoms

- Small, dark brown to black dots on leaves, petioles and stems.
- Leaf spots enlarge, coalesce forming irregular blotches.
- Infected leaves turn yellow, wither and drop.
- Black tissue may appear near base of stems.
- Crown and root rot may occur.
- Infected seed pods and seed may discolour and shrivel.

Hosts

All legumes and specially alfalfa.

Geographic Distribution

Cosmopolitan.

Biology and Transmission

Favoured by cool and wet conditions; overwinters on dead stems and leaves or in crowns and roots; spores released during periods of wet cool weather; spores spread by rain splash, wind-blown or carried by insects; and new shoots exposed through infected residue.

Detection/Indexing Method at ICARDA

- PDA.

Treatment/Control

- Use certified seed; use disease resistant cultivars; crop rotation; and crop rotation; and spring burning when severe.
- Application of benomyl seed treatments (0.1 and 0.5% w/w) resulted in only a 4-5 week delay in the onset of Phoma black stem symptoms.

Procedure followed at the Centers in Case of Positive Test

- Seed treatment.

Fusarium Wilt and Fusarium Root Rot

- Scientific names: Fusarium oxysporum, F. moniliforme, F. equiseti, F. tricinctum and F. solani. [anamorphs].

- Other scientific names: Gibberella avenacea, Gibberella moniliformis, Gibberella intricans, Gibberella tricincta, Gibberella zeae.[Teleomorphs].

Significance

The diseases cause up to 100% yield loss under heavy infestation, depending on relative humidity, soil moisture, and soil temperature.

Symptoms

The disease appears in the field in patches at both seedling and adult stages. Seedling wilt characterized by sudden drooping followed by drying of leaves and whole seedling, and apparently healthy looking roots with reduced proliferation.

The adult wilt symptoms first appear at flowering to late pod-filling stage and are characterized by sudden drooping of top leaflets of the affected plant, leaflets closure without premature shedding, dull green foliage color followed by wilting of whole plant or individual branches, apparently healthy looking roots system with a slight reduction in the lateral roots which are as difficult to pull out as the healthy plant, and no internal discoloration of the vascular system in most cases.

Hosts

Widespread.

Geographic Distribution

Worldwide.

Biology and Transmission

If a field becomes infested with either of the wilt fungi, it remains so indefinitely. The causal fungi can be carried from one field to another on farm equipment, on lentil refuse, and in wind- or waterborne soil. They may be introduced into new areas on the seed. Once introduced into a field it may take two years or more for the fungi to increase in numbers where an appreciable amount of disease is evident. A soil temperature of (23° to 27 °C) is most favorable for Fusarium or true wilt, and a slightly higher optimum temperature (27 °C) for near wilt. Race 5 of the near wilt fungus, not found in Illinois, infects at lower temperatures. The wilt-producing fungi are soil inhabitants that penetrate the lentil plant through the root hairs and fibrous roots. They grow upward through the stem, often well into the upper branches, in the water conducting tissue (xylem).

This process interferes with the passage of water from the roots to the stems, leaves, and pods resulting in yellowing, dwarfing, and wilting of plants. The Fusarium fungi do not reproduce on living plants but produce large numbers of microscopic spores (microconidia, macroconidia, and chlamydospores) in and on dead stems and roots. The spores are splashed or blown about within fields. The spores germinate and the resulting hyphae penetrate the host. The fungi survive in soil for 10 years or longer, in the absence of a lentil crop, as thick-walled chlamydospores. Survival is related to the association of the Fusarium fungi with the roots of nonhost crops. The fungi are also capable of infecting seeds.

Detection/Indexing Method at ICARDA

- Blotter Test.

Treatment/Control

- Choose resistant varieties when available. Remove stricken growth and sterilize clippers (one part bleach to 4 parts water) between cuts.
- Control garden insects, such as cucumber beetles, which are known to spread the disease.
- Remove all weeds from the garden (many weed species host the disease). The biological fungicide Mycostop will control wilt caused by Fusarium. If the disease persists, it is best to remove the entire plant and solarize the soil before planting again.
- To solarize the soil, you must leave a clear plastic tarp on the soil surface for 4-6 weeks during the hottest part of the year. Soil solarization will reduce or eliminate many soil inhabiting pests, including nematodes, fungi, insects, weeds and weed seeds.

Procedure in Case of Positive Test

- Rotation, seed treatment.

Fusarium

Stem Rot, Sclerotinia Crown Rot and Root

- Scientific name: Sclerotinia sclerotiorum (Lib.).
- Other scientific name: Hymenoscyphus sclerotiorum(Lib.).

Significance

Sclerotinia sclerotiorum infects nearly 400 plant species and causes economic damage to a wide range of crops.

Symptoms

Signs and symptoms caused by S. sclerotiorum vary depending on the host. However, the most obvious sign is the appearance of white fluffy mycelial growth that will later produce sclerotia. The fungus infects and produces mycelium at the base of the plant that will eventually move up the stem causing it to rot.

Sclerotinia sclerotiorum on seedlings (collar rot) produces small black-brown lesions at the base of the stem. These will eventually cause rapid destruction of the seedlings as the fungus moves up the stem and into the leaves. Fruits such as cucumber, squash, and eggplant are also susceptible to S. sclerotiorum. In this case the fungus causes a wet rot that will eventually engulf the entire fruit. White mycelium can be seen both externally and within the infected fruit.

Hosts

Sclerotinia sclerotiorum is among the most nonspecific, omnivorous, and successful of plant pathogens. Plants susceptible to this pathogen encompass 64 families, 225 genera, and 361 species.

Geographic Distribution

Sclerotinia sclerotiorum has a world-wide distribution; however it is most prevalent in cool moist regions.

Biology and Transmission

Approximately 90% of the life cycle of Sclerotinia species is spent in soil as sclerotia. At certain times of the year, depending on the inherent nature of the fungus and various environmental factors, the sclerotia germinate and form either mycelium which can infect a host, or an apothecium. Infection of host plants by mycelium can occur at or beneath the soil-line. Sclerotia germinate in the presence of exogenous nutrients and produce hyphae which invade nonliving organic matter, forming mycelium which then infects living host tissues. Penetration of the host cuticle is achieved by mechanical pressure. Mycelium infection is unlikely to infect plants located more than 2 cm from a sclerotium.

Apothecium produced by Sclerotium releases ascospores that disseminate by air. Generally, 10-20 °C is the optimum temperature range for carpogenic germination. Under favorable conditions, including adequate moisture, ascospores germinate within 3-6 hours of release. In one study, apothecia collected from the field discharged ascospores continuously for seven days in the laboratory. Ascospores infect nonliving host tissues, germinate, and inundate the nonliving plant part with mycelium. Then, the fungus invades healthy plant tissues with mycelium. After the plant or plant part dies, sclerotia are formed either on or in the plant. The sclerotia return to the soil for a "resting" period (which can be weeks or years) before they become active, which requires the appropriate environmental conditions.

Sclerotia are the structures which allow species of Sclerotinia to survive for long periods of time under adverse conditions. The black, melanized rind appears to act as a protectant from invasion by microorganisms. Soil temperatures, pH, and moisture appear to have little direct effect on their survival, though the combination of high temperatures and high moisture appears to encourage the degradation of sclerotia near the soil surface.

Air-borne ascospores are the most important means of spread. Mycelium from sclerotia can also cause infection, but usually this mode of infection stays within an area. Moving contaminated soil (on farm equipment, shoes, infected seedlings) and fertilizing with manure from animals fed infected plant debris, are two common ways of spreading sclerotia or mycelium from place to place. Irrigation also has been shown to be involved in the spread of Sclerotia spp. from field to field. Irrigation transported sclerotia remained viable for at least 10-21 days in flowing water. In addition, seed may be an infective source, either from contaminating sclerotia or internal mycelium. However, support for this method of transmission is questionable.

Detection/Indexing Method at ICARDA

- PDA , blotter test.

Treatment/Control

- There are essentially three methods of controlling Sclerotinia sclerotiorum; chemical, cultural, and biological control. Chemical and cultural control is most widely used, while biological control is used to a lesser degree. The only chemical control that is employed to control S. sclerotiorum is fumigation. The soil is fumigated to reduce the amount of inoculum present. Cultural practices are often used to prevent the spread of the pathogen from one host to another by killing the pathogen or by providing unfavorable conditions for the pathogen to develop. Sanitation, proper ventilation, and adequate spacing of plants are all effective cultural practices used by growers in controlling S. sclerotiorum.

Although biological control is used less extensively than chemical and cultural methods, some biological control agents have been labelled. Many antagonistic fungi such as Coniothyrium minitans, Gliocladium roseum, Gliocladium virens, Sporodesmium sclerotivorum, and Trichoderma viride have been shown to provide effective control of S. sclerotiorum. These mycoparasitic fungi inhibit the growth of S. sclerotiorum by destroying existing sclerotia while inhibiting the formation of new sclerotia. All three methods for the control of S. sclerotiorum are important to long term health and viability of the plants affected by this pathogen. Addition of compost and organic fertilizers can decrease disease levels. Fungicides such as methyl thiophanate, PCNB (pentachloronitrobenzene) and chlorothalonil are also effective. Consult label directions and carefully follow instructions for application.

- Removing the debris after harvest will reduce the amount of inoculum in the soil. In some cases, alternation with a nonsusceptible crop also reduces inoculum level.

Procedure in Case of Positive Test

- Rotation and seed treatments.

Sclerotinia disease.

Stemphylium Leaf Blight

Scientific names: Stemphylium botryosum Rotem,Y. Cohen & I. Wahl, (1966) [Anamorph]: Pleospora herbarum(Pers.) Rabenh., (1854) [Teleomorph].

Significance

It is reported to be seedborne.

Symptoms

Spot-causing fungal disease occurring in leaf. The lesions are at first water-soaked small spots and then become blackish brown lesion divided by leaf veins. The lesions often extend from the leaf rim, fuse with the adjacent lesion, and become large. The leaf shrinks and withers in results. The diseased leaf defoliates at the early stage. The species of the causal organism is same with leaf spot fungi of alfalfa, but the pathogenicity is differentiated.

Hosts

On a wide range of hosts including apple, ash, broad bean, clover, endive, gladiolus, gramineae, lettuce, lupin, muskmelon, onion, Onobrychis, Medicago sativa, mangold, tomato, Trifolium, Vicia.

Geographic Distribution

Worldwide; very common in temperate and sub-tropical regions.

Biology and Transmission

- Survival: The fungi are carried-over on infected plant debris (leaves and stems), seed and in the soil.
- Environmental conditions: Stemphylium leaf spot may develop under warm moist conditions.
- Dispersal: Spores of these fungi are spread by wind and rain splash. Generally air-borne as ascospores or conidia. These penetrate the leaf or petiole via stomata. May also occur on seeds and in soil.

Detection/Indexing Method at ICARDA

- Blotter test, agar plate method.

Treatment/Control

- Seed treatment. These diseases are usually not a problem after the end of the rainy season except in Valleys where they can cause damage up to harvest under conditions of high humidity and heavy night time dew. (ICARDA).

Procedure followed in Case of Positive Test

- Seed treatment.

Damping-Off

Scientific names: Rhizoctonia solaniJ.G. Kühn 1858 [Anamorph], Thanatephorus cucumeris(A.B. Frank) Donk 1956 [Teleomorph].

Significance

Not Significant.

Symptoms

R. solani primarily attacks below ground plant parts such as the seeds, hypocotyls, and roots, but is also capable of infecting above ground plant parts (e.g. pods, fruits, leaves and stems). The most common symptom of Rhizoctonia disease is referred to as "damping-off" characterized by non-germination of severely infected seed whereas infected seedlings can be killed either before or after they emerge from the soil. Infected seedlings not killed by the fungus often have cankers, which are reddish-brown lesions on stems and roots. In addition to attacking below ground plant parts, the fungus will occasionally infect fruit and leaf tissue located near or on the soil surface. This type of disease often occurs because the mycelium and/or sclerotia of the fungus are close to or splashed on the plant tissue.

Although most Rhizoctonia diseases are initiated by mycelium and/or sclerotia, several important diseases of beans, result from basidiospore infection. These basidiospores also serve as a source for rapid and long distance dispersal of the fungus. The basidiospores germinate to produce hyphae that infect leaves during periods of high relative humidity and periods of extended wet weather. Under these conditions, basidiospores can often be observed on the base of stems near the soil surface or on the underside of leaves in the plant canopy.

The pathogen is transported in infested soil or through movement of diseased plants or bean pods. Potential for seed-borne inoculum also exists. Although basidiospores are wind-borne, their role in initiating disease has not been considered important except for foliar diseases (web blight) in high humidity.

Hosts

R. solani is a very common soilborne pathogen with a great diversity of host plants.

Geographic Distribution

Common throughout the world.

Biology and Transmission

Rhizoctonia solani produces thread-like growth called hyphae ; large masses of hyphae are referred to as mycelium. The hyphae of Rhizoctonia solani have the following characteristics, some shade of brown, a special type of cross wall within the hyphae, called a dolipore septum, each cell is multinucleate (has many nuclei) rather than binucleate, branches that are produced at right angles, no asexual spores are formed by the mycelium.

In general, the growth rate of Rhizoctonia solani is very rapid and a typical isolate can grow across a 90 mm petri plate in three days. Small, oval cells produced in branched chains or clusters are formed. These are called monilioid cells and have slightly thicker walls than the mycelium. Large aggregates of these cells are called sclerotia which are black to brown and 3 to 5 mm long.

The hyphae of Rhizoctonia solani have many nuclei (commonly 4 to 8) per cell. This distinguishes it from similar fungi that have only 2 nuclei per cell. Those fungi with hyphal characteristics similar to Rhizoctonia solani, but with only 2 nuclei per cell, are called binucleate types and are generally non-pathogenic. These Rhizoctonia solani-like fungi are saprophytic, do not cause disease, and feed on dead organic matter.

Following invasion of the host by Rhizoctonia solani, sexual spores are formed on specialized structures called basidia. Four spores are produced on each basidium. Basidia are formed when the environment is moist and sufficient growth of the fungus has occurred. Basidiospores are wind-dispersed and germinate with moisture. Each basidiospore has a single nucleus. The hyphae produced by germinating spores will fuse (anastamose), forming new hyphae with a mixture of different types of nuclei.

Observations of basidiospores on diseased host tissue are not common. Several methods have been developed to produce basidiospores in pure culture but it remains difficult.

In recent years, the recognition of different anastomosis groups has helped to elucidate the complexity of this large fungal group. Having no asexual spores and only the few morphological characteristics to define "Rhizoctonia solani isolates", many fungi were placed into this species. However, it is now recognized that Rhizoctonia solani has multinucleate cells and other related fungi (with 2 nuclei per cell), the binucleates, do not belong in that species. Further, when isolates are paired, only related mycelium will fuse and thus anastomosis groups have been defined by a set of tester isolates. A set for Rhizoctonia solani isolates and another set for binucleate Rhizoctonia solani-like fungi are also available.

Detection/Indexing Method at ICARDA

- Agar Plate Test.

Treatment/Control

- Obtain high quality seeds and avoid seeds that may be contaminated with the pathogen. In fields known to have Rhizoctonia solani, prepare planting areas to increase drainage and prevent water accumulation. Set plants with adequate spacing to avoid crowding and formation of high humidity in the foliage.
- Addition of compost and organic fertilizers can decrease disease levels. Fungicides such as methyl thiophanate, PCNB (pentachloronitrobenzene) and

chlorothalonil are also effective. Consult label directions and carefully follow instructions for application.

- Removing the debris after harvest will reduce the amount of inoculum in the soil. In some cases, alternation with a nonsusceptible crop also reduces inoculum level.

Procedure followed in Case of Positive Test

- Rotation, seed treatment.

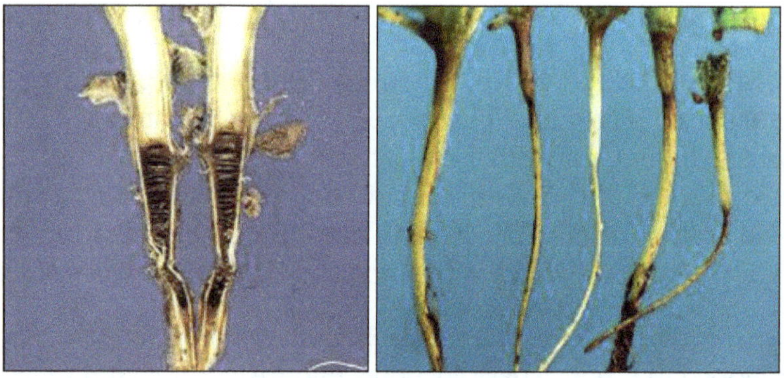

Rhizoctonia root disease.

Diseases in Wheat

Powdery Mildew

Disease Symptoms:

- Powdery mildew can easily be diagnosed by the white, powdery patches that form on the upper surface of leaves and stem.
- Greyish white powdery growth appears on the leaf, sheath, stem and floral parts.
- Powdery growth later become black lesion and cause drying of leaves and other parts.

Survival and Spread:

- Fungus remains in high hills during summers in infected plant debris as dormant mycelium and asci. Primary spread is by the asciospores and secondary spread through airborne conidia.

Favourable Conditions:

- The disease infects plants during periods of high humidity (not necessarily rain) and cool to moderate temperatures (20-21 °C).

Loose Smut

Disease Symptoms:

- It is a seed borne disease; infection occurs during Loose Smut flowering through wind-borne spores.
- The infection remains dormant inside the otherwise healthy looking seed but the plants grown from such seeds bear infected inflorescence.
- At this time, infected heads emerge earlier than normal heads. The entire inflorescence is commonly affected and appears as a mass of olive-black spores, initially covered by a thin gray membrane.
- Once the membrane ruptures, the head appears powdery.

Survival and Spread:

- The disease is internally seed borne, where pathogen infects the embryo in the seed.
- Primary infection occurs by sowing infected seeds.

Favourable Conditions:

- Infection is favored by cool, humid conditions during flowering period of the host plant.

Brown Rust

Disease Symptoms:

- The most common site for symptoms is on upper leaf blades, however, sheaths, glumes and awns may occasionally become infected and exhibit symptoms.
- The pustules are circular or slightly elliptical, smaller than those of stem rust, usually do not coalesce, and contain masses of orange to orange-brown Urediospores.

Survival and Spread:

- Pathogen over-summers in low and mid altitudes of Himalayas and Nilgiris. Primary infections develop from wind deposited urediospores in eastern Indo-gangetic plains in middle of January where it multiplies and moves westwards by March.
- Alternate host is Thalictrum sp.

Favourable Conditions:

- Temperatures of 20-25 °C with free moisture (rain or dew) cause epidemics. Severe infection causes upto 30 percent yield losses.

Stripe Rust/Yellow Rust

Disease Symptoms:

- Mainly occur on leaves than the leaf sheaths and stem. Bright yellow pustules (Uredia) appear on leaves at early stage of crop and pustules are arranged in linear rows as stripes.
- The stripes are yellow to orange yellow. The teliospores are also arranged in long stripes and are dull black in colour.
- The pustules of stripe rust, which, contain yellow to orange-yellow urediospores, usually form narrow stripes on the leaves.
- Pustules also can be found on leaf sheaths, necks, and glumes.

Survival and Spread:

- The inoculum survives in the form of uredospores/teliospores in the northern hills during off season on self-sown crop or volunteer hosts, which provide an excellent source of inoculums and primary spread occur through uredospores from hills.

Favourable Conditions:

- Disease is prominent when temperature is 10-20 °C and high humidity.

Black Rust

Disease Symptoms:

- Symptoms are produced on almost all aerial parts of the wheat plant but are most common on stem, leaf sheaths and upper and lower leaf surfaces.
- Pustules (containing masses of urediospores) are dark reddish brown - occur on both sides of the leaves, on the stems, and on the spikes.
- Pustules are usually separate and scattered, heavy infections -coalesce.
- Prior to pustule formation, "flecks" may appear. Before the spore masses break through the epidermis, the infection sites feel rough to the touch.
- As the spore masses break through, the surface tissues take on a ragged and torn appearance.

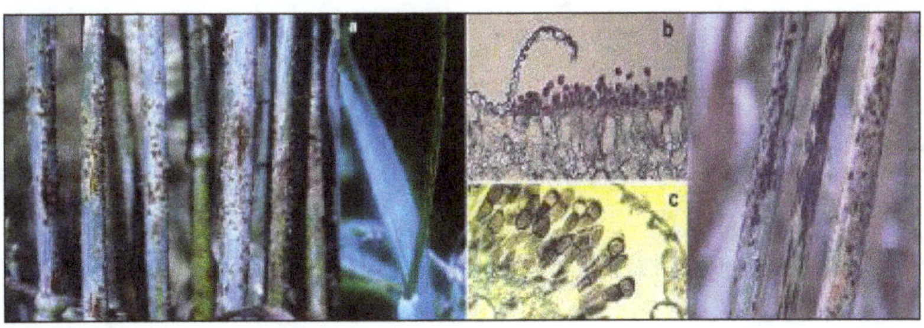

Survival and Spread:

- Both survive on stubbles and volunteer crops, alternate host: Berberis spp. and primary spread occur through uredospore from southern hills.

Favourable Conditions:

- Moisture and temperature above 20° C favours the development of disease.

Flag Smut

Disease Symptoms:

- Symptoms can be seen on stem, clum and leaves from late seedling stage to maturity.
- The seedling infection leads to twisting and drooping of leaves followed by withering.
- Grey to grayish black sori occurs on leaf blade and sheath. The sorus contains black powdery mass of spores.

Survival and Spread:

- The disease is seed and soil borne. Smut spores are viable for more than 10 years.
- Primary infection occurs by sowing infected seeds or by resting spores present in the soil.

Favourable Conditions:

- Temperature of 18-24 °C.
- Relative humidity 65% and above.

Hill Bunt or Stinking Smut

Disease Symptoms:

- The fungus attacks seedling of 8-10 days old and become systemic and grows along the tip of shoot.
- At the time of flowering hyphae concentrate in the inflorescence and spikelets

and transforming the ovary into smut sorus of dark green color with masses of chlamydospores.

- The diseased plants mature earlier and all the spikelets are affected.

Survival and Spread:

- The pathogen survives in seeds and sowing such seeds are source of primary infection.

Favourable Conditions:

- Temperature of 18-20 °C.
- High soil moisture.

Karnal Bunt

Disease Symptoms:

- Symptoms of Karnal bunt are often difficult to distinguish in the field due to the fact that incidence of infected kernels on a given head is low.
- There may be some spreading of the glumes due to sorus production but it is not as extensive as that observed with common bunt.
- Symptoms are most readily detected on seed after harvest.

Survival and Spread:

- The disease is seeds borne and sowing of infected seeds is the source of primary infection.

Fabourable Conditions:

- Temperature of 18-20 °C.
- High soil moisture.

Leaf Blight

Disease Symptoms:

- Reddish brown oval spots appear on young seedlings with bright yellow margin. In severe cases, several spots coalesce to cause drying of leaves.
- It is a complex disease, having association of A. triticina, B. sorokiniana and A. alternate.

Survival and Spread:

- Primary spread is by externally seed-borne and soil borne conidia. Secondary spread by air-borne conidia.

Favourable Conditions:

- Temperature of 25 °C and high relative humidity.

Foot Rot

Disease Symptoms:

- The disease mainly occurs in seedlings and roots and rootlets become brown in colour.
- Seedlings become pale green and have stunted growth.
- Fungus produces sporangia and zoospores and oospores.

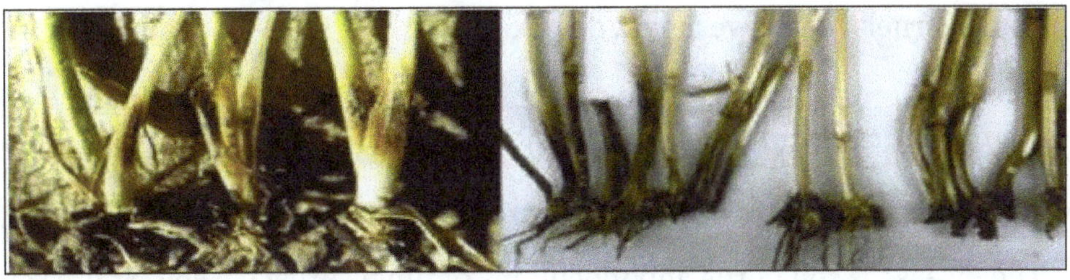

Survival and Spread:

- The disease is soil borne, pathogens survives in soil.
- Primary spread occurs through soil and irrigation water.

Favourable Conditions:

- Wet weather and high rainfall.

Head Scab/Fusarium Leaf Blotch (Snow Mold)

Disease Symptoms of Leaf Blotch:

- The blotching caused by this organism becomes evident on leaves at about late-joint to early-boot growth stage.
- Young lesions occur as oval to elliptical, greyish green mottled areas, usually located where the leaf bends. The lesions enlarge rapidly, developing into large, "eyespot" blotches with bleached or light grey centers; the leaves tend to split or shred, beginning at the centers of the lesions.

Disease Symptoms of Head Scab:

- The fungus also can cause head scab.
- Symptoms of Fusarium head blight include tan or light brown lesions encompassing one or more spikelets. Some diseased spikelets may have a dark brown discoloration at the base and an orange fungal mass along the lower portion of the glume.
- Grain from plants infected by Fusarium head blight is often shriveled and has a white chalky appearance.

- Some kernels may have a pink discoloration.
- Infected florets (especially the outer glumes) become slightly darkened and oily in appearance.

Survival and Spread:

- The disease is soil borne and inoculums of fungi survive in soil. Spores are produced on crop debris left on or near the soil surface. These spores are transmitted to leaves by the wind or by splashing rain.

Favourable Conditions:

- Disease development is favored by cool, moist weather.

Helminthosporium Leaf Blotch (Spot Blotch)

Disease Symptoms:

- Lesions caused by this disease are elongated to oval in shape and are generally a dark brown color.
- As lesions mature, the centers often turn a light brown to tan color, surrounded by an irregular dark brown ring (21 on leaf; 22 on spike).
- Primary infections tend to be on the lower leaves, beginning as chlorotic flecks or spots. These infection sites enlarge, turn dark brown, and often coalesce. When the disease is severe, affected leaves or leaf sheaths may die prematurely.

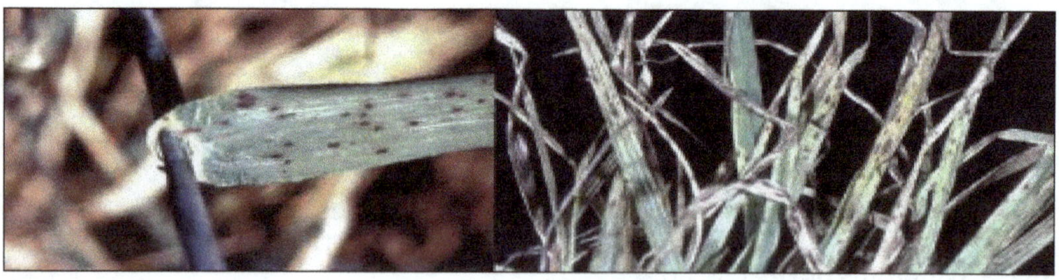

Survival and Spread:

- The Disease is seed as well as soil borne and inoculums present in the seeds and soil are the source of primary infection.

Favourable Conditions:

- Disease is prevalent in more humid and higher rainfall areas.

Seedling Blight

Disease Symptoms:

- This symptom is seen when ears become infected during the early flowering stages. Later infections may result in infection of the grain but without obvious bleaching of the ears.
- Fusarium lesions often begin in the leaf sheath at the stem base where crown roots split the leaf sheath when emerging. This infection can then spread up the leaf sheath causing long dark brown streaks at the stem base.
- The ear blight phase of the disease can cause yield loss but is most important as it can result in mycotoxin production in the grain.

Survival and Spread:

- The most important source of Fusarium for wheat crops is the seed but the fungus can also survive on debris in the soil.

Favourable Conditions:

- Relatively high soil moisture and soil temperature are favourable for the infection.

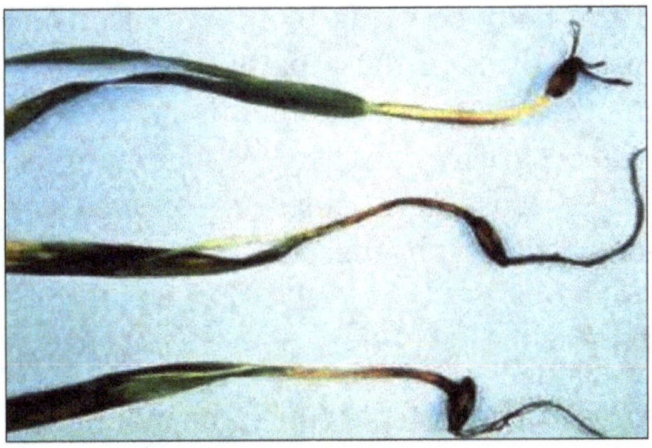

Chapter 3

Insecticides

The substances which are used to kill insects are known as insecticides. They can be grouped into three categories: inorganic insecticides, organochloride insecticides and organophosphate insecticides. All these types of insecticides are thoroughly discussed in this chapter.

Insecticide

Insecticide, any toxic substance that is used to kill insects. Such substances are used primarily to control pests that infest cultivated plants or to eliminate disease-carrying insects in specific areas.

Insecticides can be classified in any of several ways, on the basis of their chemistry, their toxicological action, or their mode of penetration.They are classified according to whether they take effect upon ingestion (stomach poisons), inhalation (fumigants), or upon penetration of the body covering (contact poisons). Most synthetic organic insecticides penetrate by all three of these pathways, however, and hence are better distinguished from each other by their basic chemistry. Besides the synthetic organics, some organic compounds occurring naturally in plants are useful insecticides, as are some inorganic compounds. Most insecticides are sprayed or dusted onto plants and other surfaces traversed or fed upon by insects.

Modes of Penetration

Stomach poisons are toxic only if ingested through the mouth and are most useful against those insects that have biting or chewing mouth parts, such as caterpillars, beetles, and grasshoppers. The chief stomach poisons are the arsenicals—e.g., Paris green (copper acetoarsenite), lead arsenate, and calcium arsenate; and the fluorine compounds, among them sodium fluoride and cryolite. They are applied as sprays or dusts onto the leaves and stems of plants eaten by the target insects. Stomach poisons have gradually been replaced by synthetic organic insecticides, which are less dangerous to humans and other mammals.

Contact poisons penetrate the skin of the pest and are used against those arthropods, such as aphids, that pierce the surface of a plant and suck out the juices. The contact insecticides can be divided into two main groups: naturally occurring compounds and synthetic organic ones. The naturally occurring contact insecticides include nicotine, developed from tobacco; pyrethrum, obtained from flowers of *Chrysanthemum cinerariaefolium;* rotenone, from the roots of *Derris* species and related plants; and oils, from petroleum. Though these compounds were originally derived mainly from plant extracts, the toxic agents of some of them (e.g., pyrethrins) have been synthesized. Natural insecticides are usually short-lived on plants and cannot provide protection against prolonged invasions. Except for pyrethrum, they have largely been replaced by newer synthetic organic insecticides.

Fumigants are toxic compounds that enter the respiratory system of the insect through its spiracles, or breathing openings. They include such chemicals as hydrogen cyanide, naphthalene, nicotine, and methyl bromide and are used mainly for killing insect pests of stored products or for fumigating nursery stock.

Type of Activity

Systemic insecticides become incorporated and distributed systemically throughout the whole plant. When insects feed on the plant, they ingest the insecticide. Systemic insecticides produced by transgenicplants are called plant-incorporated protectants (PIPs). For instance, a gene that codes for a specific *Bacillus thuringiensis* biocidal protein was introduced into corn (maize) and other species. The plant manufactures the protein, which kills the insect when consumed.

Contact insecticides are toxic to insects upon direct contact. These can be inorganic insecticides, which are metals and include arsenates, copper and fluorine compounds, which are less commonly used, and the commonly used sulfur. Contact insecticides can be organic insecticides, i.e. organic chemical compounds, synthetically produced, and comprising the largest numbers of pesticides used today. Or they can be natural compounds like pyrethrum, neem oil etc. Contact insecticides usually have no residual activity.

Efficacy can be related to the quality of pesticide application, with small droplets, such as aerosols often improving performance.

Biological Pesticides

Many organic compounds are produced by plants for the purpose of defending the host plant from predation. A trivial case is tree rosin, which is a natural insecticide. Specifically, the production of oleoresin by conifer species is a component of the defense response against insect attack and fungal pathogen infection. Many fragrances, e.g. oil of wintergreen, are in fact antifeedants.

Four extracts of plants are in commercial use: pyrethrum, rotenone, neem oil, and various essential oils.

Other Biological Approaches

Plant-incorporated Protectants

Transgenic crops that act as insecticides began in 1996 with a genetically modified potato that produced the Cry protein, derived from the bacterium Bacillus thuringiensis, which is toxic to beetle larvae such as the Colorado potato beetle. The technique has been expanded to include the use of RNA interference RNAi that fatally silences crucial insect genes. RNAi likely evolved as a defense against viruses. Midgut cells in many larvae take up the molecules and help spread the signal. The technology can target only insects that have the silenced sequence, as was demonstrated when a particular RNAi affected only one of four fruit fly species. The technique is expected to replace many other insecticides, which are losing effectiveness due to the spread of pesticide resistance.

Enzymes

Many plants exude substances to repel insects. Premier examples are substances activated by the enzyme myrosinase. This enzyme converts glucosinolates to various compounds that are toxic to herbivorousinsects. One product of this enzyme is allyl isothiocyanate, the pungent ingredient in horseradish sauces.

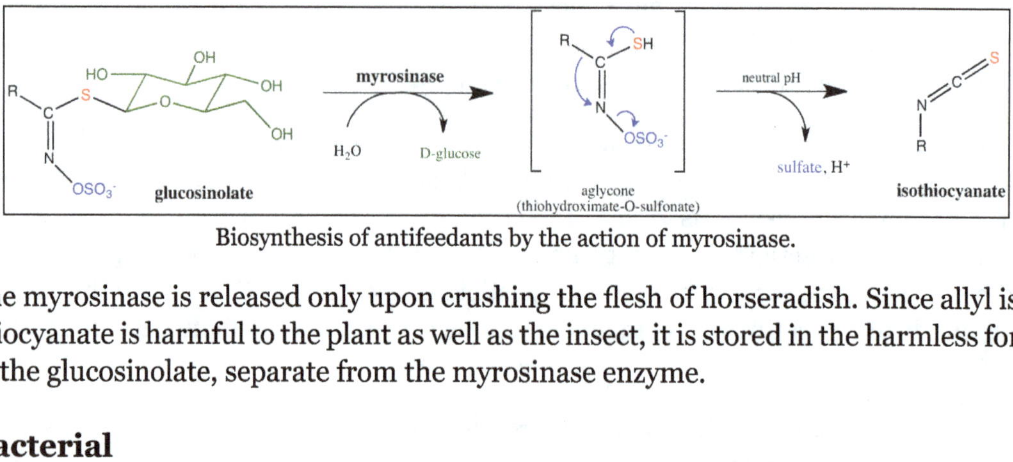

Biosynthesis of antifeedants by the action of myrosinase.

The myrosinase is released only upon crushing the flesh of horseradish. Since allyl isothiocyanate is harmful to the plant as well as the insect, it is stored in the harmless form of the glucosinolate, separate from the myrosinase enzyme.

Bacterial

Bacillus thuringiensis is a bacterial disease that affects Lepidopterans and some other insects. Toxins produced by strains of this bacterium are used as a larvicide against caterpillars, beetles, and mosquitoes. Toxins from *Saccharopolyspora spinosa* are isolated from fermentations and sold as Spinosad. Because these toxins have little effect on other organisms, they are considered more environmentally friendly than synthetic pesticides. The toxin from *B. thuringiensis* (Bt toxin) has been incorporated directly into plants through the use of genetic engineering. Other biological insecticides include products based on entomopathogenic fungi (e.g., *Beauveria bassiana, Metarhizium anisopliae*), nematodes (e.g., *Steinernema feltiae*) and viruses (e.g., *Cydia pomonella* granulovirus).

Synthetic Insecticide and Natural Insecticides

A major emphasis of organic chemistry is the development of chemical tools to enhance agricultural productivity. Insecticides represent a major area of emphasis. Many of the major insecticides are inspired by biological analogues. Many others are completely alien to nature.

Organochlorides

The best known organochloride, DDT, was created by Swiss scientist Paul Müller. For this discovery, he was awarded the 1948 Nobel Prize for Physiology or Medicine. DDT was introduced in 1944. It functions by opening sodium channels in the insect›s nerve cells. The contemporaneous rise of the chemical industry facilitated large-scale production of DDT and related chlorinated hydrocarbons.

Organophosphates and Carbamates

Organophosphates are another large class of contact insecticides. These also target the insect›s nervous system. Organophosphates interfere with the enzymes acetylcholinesterase and other cholinesterases, disrupting nerve impulses and killing or disabling the insect. Organophosphate insecticides and chemical warfare nerve agents (such as sarin, tabun, soman, and VX) work in the same way. Organophosphates have a cumulative toxic effect to wildlife, so multiple exposures to the chemicals amplifies the toxicity. In the US, organophosphate use declined with the rise of substitutes.

Carbamate insecticides have similar mechanisms to organophosphates, but have a much shorter duration of action and are somewhat less toxic.

Pyrethroids

Pyrethroid pesticides mimic the insecticidal activity of the natural compound pyrethrum, the biopesticide found in pyrethrins. These compounds are nonpersistent sodium channel modulators and are less toxic than organophosphates and carbamates. Compounds in this group are often applied against household pests.

Neonicotinoids

Neonicotinoids are synthetic analogues of the natural insecticide nicotine (with much lower acute mammalian toxicity and greater field persistence). These chemicals are acetylcholine receptor agonists. They are broad-spectrum systemic insecticides, with rapid action (minutes-hours). They are applied as sprays, drenches, seed and soil treatments. Treated insects exhibit leg tremors, rapid wing motion, styletwithdrawal (aphids), disoriented movement, paralysis and death. Imidacloprid may be the most common. It has recently come under scrutiny for allegedly pernicious effects on honeybees and its potential to increase the susceptibility of rice to planthopper attacks.

Ryanoids

Ryanoids are synthetic analogues with the same mode of action as ryanodine, a naturally occurring insecticide extracted from *Ryania speciosa* (Flacourtiaceae). They bind to calcium channels in cardiac and skeletal muscle, blocking nerve transmission. The first insecticide from this class to be registered was Rynaxypyr, generic name chlorantraniliprole.

Insect Growth Regulators

Insect growth regulator (IGR) is a term coined to include insect hormone mimics and an earlier class of chemicals, the benzoylphenyl ureas, which inhibit chitin (exoskeleton) biosynthesis in insectsDiflubenzuron is a member of the latter class, used primarily to control caterpillars that are pests. The most successful insecticides in this class are the juvenoids (juvenile hormone analogues). Of these, methoprene is most widely used. It has no observable acute toxicity in rats and is approved by World Health Organization (WHO) for use in drinking water cisterns to combat malaria. Most of its uses are to combat insects where the adult is the pest, including mosquitoes, several fly species, and fleas. Two very similar products, hydroprene and kinoprene, are used for controlling species such as cockroaches and white flies. Methoprene was registered with the EPA in 1975. Virtually no reports of resistance have been filed. A more recent type of IGR is the ecdysone agonist tebufenozide (MIMIC), which is used in forestryand other applications for control of caterpillars, which are far more sensitive to its hormonal effects than other insect orders.

Examples

- Organochlorides.
- Organophosphates.
- Carbamates.
- Pyrethroids.
- Neonicotinoids.
- Ryanoids.
- Insect growth regulators.
- Derived from plants or microbes.

Some insecticides, such as insecticidal soaps, work through non-toxic methods, but others contain chemicals that are harmful to other forms of life. Insecticides are also ineffective when used excessively because pests build up resistance. Non-chemical control methods don't have these disadvantages.

Non-Target Organisms

Insecticides can affect many more organisms than the ones they're intended to kill. Insecticides are not without risks to the humans who use them. And when insecticides enter water sources through runoff, drift, leaching or accidental release, they kill fish and other aquatic wildlife. Birds die from drinking contaminated water and eating affected insects, as well as from accidental exposure to sprays. The long-term effects of insecticides on non-target organisms are unknown. Some remain in the bodies of animals that eat them and travel up the food chain, becoming more concentrated until finally affecting top predators. Some insecticides, including DDT, were largely banned in the United States because they were found to affect the reproductive abilities of predatory birds.

Beneficial Insects

Insecticides can kill insects that help gardeners and farmers. Honeybees are essential for crop pollination as well as making honey, yet they and other pollinators are killed by widely-used insecticides. Insecticides also kill ladybugs, predatory wasps, lacewings, hoverflies, ground beetles, spiders and other insects that eat pests. Gardeners sometimes make infestations worse by killing a pest's natural predators. The wider roles insects and microorganisms that feed on pests play in agriculture and the environment are probably underestimated, states the Institute of Food and Agricultural Sciences, University of Florida.

Resistance

Repeated exposure to insecticides builds up resistance in insects, until finally the insecticide has little or no effect. Every insect population produces some individuals with a genetic makeup that allows them to withstand the insecticide's effects. These individuals survive the spraying and pass on their protective genetic makeup to their offspring. Over time, the population of resistant individuals increases until it forms the majority. Insects reproduce quickly, often as frequently as a new generation every three or four weeks, so resistance can build up rapidly. Frequent insecticide applications make the problem worse.

Minimizing Effects

Using selective, non-persistent insecticides and taking precautions when spraying help minimize the disadvantages of insecticides. Selective insecticides kill only a narrow range of organisms, and non-persistent insecticides break down quickly in the environment. Use ready-mixed products to avoid having to contaminate other containers with pesticide, and spray only where the pest attack is most severe. Don't spray on windy days or when rain is forecast, and spray in the early morning to have the least impact on bees. Non-chemical pest control methods include optimizing growing conditions, growing resistant varieties, using insect traps or barriers, hosing plants down and growing plants that attract beneficial insects.

Inorganic Insecticides

Inorganic Insecticides are insecticides that are either man-made or based on synthetic formulas like carbamates or pyrethroids. These are made from heavy metals and arsenic compounds such as boric acid and silica gel. Inorganic types of insecticides are highly effective but use them carefully and read the applying conditions properly. There are several types of inorganic insecticides.

Systematic Insecticides are put in the soil and allowed to be absorbed by plant roots. These insecticides move up to the leaves, branches, fruits and twigs of the plant and protect it from being chewed upon by insects. Bactericides and larvicides are also a form of systematic insecticides that help in eradicating bacteria and larvae that destroys plants.

Contact Insecticides are the common sprays that can be used at home to eradicate insects and mosquitoes. Some of these products are also used to disinfect the home and environment and deodorise the place. These insecticides are available in the market in the form of sprays, coils, liquid vaporisers and repellents.

Ingested Insecticides are used to kill rats, rodents and cockroaches. Rodenticide is also a form of ingested insecticide. It is sort of a poison that is effective in killing rats. It is used as poisonous bait and fed to rats to get rid of them.

Herbicides are yet another kind of pesticides that are used to kill infected plants. It is specifically used to destroy parts of the plant that the insects feed on or the sections of the plant where infection has spread. Fungicides are also from the same family of herbicides; they are used to kill weeds and fungi that are responsible for destroying the agricultural yield.

Paris Green

Paris green, also called Schweinfurt green, an extremely poisonous, bright green powder that was formerly used extensively as a pigment (e.g., in wallpaper) and that is sometimes used as an insecticide or to kill plant fungi it must be used with great caution because of its poisonous nature. Chemically it is a copper acetoarsenite that may be prepared from arsenic trioxide and copper acetate.

Properties

Paris green is a blue-green solid. It is extremely toxic. It has a very bright blue-green color. It darkens in air.

Preparation

It is made by reacting copper(II) acetate with arsenic trioxide and water.

Uses

Paris green was first used to kill rats in sewers in Paris. This is where it got its name from. It was used as an insecticide. It was dropped from airplanes to stop malaria in Italy. It was used as a bright green pigment. It was painted on ships to prevent barnacles from growing. In many places, Paris green is banned because it is so toxic. It is only used in a very few places.

Safety

Paris green is extremely toxic. It used to be common and so was a common poison.

Boric Acid

Boric acid, also called orthoboric acid, boracic acid, hydrogen borate and acidum boricum, is a low-toxicity mineral with insecticidal, fungicidal, and herbicidal properties. It is the "*secret ingredient*" in many of the commercial products used for insect control. It is also a weak acid of boron, often used as a flame retardant, antiseptic, or precursor to other chemical compounds.

Its chemical formula is H_3BO_3, and exists in the form of a white powder or colourless crystals that dissolves in warm water. It does not evaporate or volatilise into the air.

Boron, the main element of boric acid, doesn't exist in nature by itself. It must be combined with other elements to create compounds. Boron is an essential micro nutrient for plants and humans. We get 1-3 milligrams of boron every day from the food we eat. Grains, nuts, fruits and vegetables are good sources of boron.

Boric acid is known as a desiccant which means it induces or sustains a state of dryness in its local vicinity. It acts as a insect killer by removing moisture from the body causing severe dehydration and eventually death. It maintains its potency for a long period of time and is highly effective in extremely small amounts.

While it is not as fast acting as synthetic pesticides, it does not pose the considerable health concerns as the synthetic pesticides and remains highly effective over a long period of time.

It is odourless and, unlike other insecticides, is non-repellent. That means that insects will continue to return to the treated areas.

Effectiveness

Boric acid is excellent for killing insects. It works as a desiccant to dehydrate the insects, eventually causing death. It also attaches to the insects' feet as they walk through it allowing it to be taken back to the nest. It is neither an attractant nor a repellent so the insects will keep coming back to treated areas.

Within 2-3 weeks you should start to see some results and it can take up to a month to spread through an entire colony.

Boron is an essential nutrient for growth and development of healthy plants. Boron compounds are used in small concentrations as micronutrients in fertilizers. When used in large concentrations they function as herbicides, algaecides and other pesticides.

Functions of boron in plants:

- Essential for maintaining a balance between sugar and starch and helps in the translocation of sugar and carbohydrates.
- Important in pollination and seed production.
- Necessary for normal cell division, nitrogen metabolism and protein formation.

Boron is an essential element for plants' development, growth, crop yielding and seed development. It helps with the transfer of water and nutrition in plants. Though plants' boron requirements are very low in quantity, their growth and crop yields are severely affected when soil is boron deficient.

Boric acid is commonly used in the agriculture industry as a source of boron for liquid fertilizers. These liquid boron fertilizers are typically about 10% boron, and are used for soil and foliar applications.

Borates in solid form can be applied directly onto soil, where in humid conditions they will readily dissolve then be taken up by the plants' roots. Typical products used for this application are granular borax pentahydrate (Etibor 48) and granular ulexite. These products can be applied directly, but are more commonly used as part of a nutrient blend containing other fertilizer components. The amount of boron fertilizer to be applied onto soil varies according to plant type, method of application, amount of rain, and the soil's lime and organic material contents.

Boron products can also be applied by spraying onto plant leaves. This benefits some plants whose intake is better through their leaves or in other cases where spraying is a better way of fertilizing. Disodium octaborate tetrahydrate (Etidot 67), which is specially made for agricultural applications, is the preferred boron product for these foliar sprays because it has much better solubility compared to the conventional boron products like borax decahydrate and borax pentahydrate, and a higher boron content than boric acid.

Other products used in fertilizer include borax decahydrate and borax pentahydrate. Because of their good solubility, sodium borates can be applied directly onto soil or sprayed onto plants. Ulexite, which is a naturally occurring calcium borate, is favored in sandy soils because of its low solubility. This helps it to remain in soil longer than sodium borates.

Boron-based herbicides are produced from borax and boric acid. They are generally mixed with sodium chloride or other chemicals.

Aluminium Phosphide

Aluminum Phosphide is a useful inorganic compound used as a pesticide. It is mainly used as a fumigant for pest control and as a source of Phosphine gas (also used for fumigation). This inorganic compound can also be used as a wide band gap or energy gap semiconductor.

It is a reducing agent, which indicates that it acts as an electron donor in a chemical reaction.

The usefulness of this compound as a source of Phosphine gas or Hydrogen Phosphide (PH3) was first discovered by a German company named Degesch. In 1958, the United States registered it as a pesticide to the Hollywood Termite Control Company, Inc.

Aluminum Phosphide Applications

It can be used as:

- Fumigant
- Insecticide
- Rodenticide
- Semiconductor

Uses of Aluminum Phosphide as Pesticide in Fumigation

Fumigation is the process of pest control for grains, soil, buildings and other different stored goods. In this case, the AlP is left to react with water and the resultant Phosphine gas is then used as the fumigant.

In case of fumigation of grains stored in silos, bins or any other similar container, 2- 4 Aluminum Phosphide tablets should be applied per cubic meter. If the grain is stored in a flat storage area, 3- 5 tablets per cubic meter will be adequate. In case of packed food or other stored goods, it will be sufficient to apply 1 to 1.5 tablets per cubic meter.

This inorganic chemical is used to control both the insects and rodents. Following are some information about its uses in the field of pest control:

Insecticide

Insecticide refers to a specific type of pesticide that is used only against insects. AlP is very popular as an insecticide as it is highly effective and very easy to use. The AlP insecticides are usually non-flammable, but it reacts with air moisture and creates PH3 gas which is very toxic and flammable. Due to this reason, these insecticides also contain suitable stabilizers, usually Ammonium Carbamate, to control the development of the poisonous gas.

Rodenticide

It indicates another type of pesticide that is specifically used against rodents. It is widely used in the US and is among the most poisonous pesticides. Aluminum Phosphide rodenticide can be very effective in driving away small mammals like rats, rabbits and moles. The rodenticide pellets are mixed with their food. Once they ingest it, the chemical reacts with the acids in the digestive system of the animal and produces PH3. As a result, the animal gets poisoned.

The use of this material as a fumigant has caused accidental deaths in many countries like Saudi Arabia and the US. The Iranian Forensic Medicine Organization has started a campaign to stop the use of this chemical as a pesticide.

Organochloride Insecticides

Organochloride pesticides are very effective at eliminating pests, especially insects. But many of these chemical products are viewed negatively by environmental activists and consumers because of one well-known and now banned organochloride pesticide: dichloro diphenyl trichoroethane, better known as DDT.

Organochloride pesticides refer to chemicals with carbon, chlorine and hydrogen components. As the U.S. Fish and Wildlife Service explained, chlorine-carbon bonds are particularly strong, which keeps these chemicals from breaking down quickly or being soluble in water. The chemical is also attracted to fat and is stored in the fatty tissue of animals that consume it. The durability of organochloride pesticides' chemical makeup is one of the reasons it is so effective as an insecticide as well as potentially harmful – it can protect crops for a long time, but also stay in an animal's system.

Along with DDT, other organochloride pesticides, such as aldrin, dieldrin, heptachlor, mirex, chlordecone and chlordane, have been banned by the U.S. Environmental Protection Agency. Europe has similarly banned many organochloride pesticides, but in both of these regions organochloride chemicals are still the active ingredients in a number of home, garden and environmental pest control products, according to the EPA. Organochloride pesticides are also enormously popular in developing countries across the globe for agricultural uses.

Testing for Organochloride Pesticides

Whether examining farm land to see if it is still filled with years-old organochloride pesticides or inspecting water for organochlorines, testing is the best way to know if you have these chemicals near you. EPA methods 8250A and 8270B can be used to test for these chemicals. 8250A can test waste, soil and water, while 8270B uses gas chromatography/mass spectrometry (GC/MS).

Although organochloride pesticides are most famous for damaging some birds' abilities to lay healthy eggs, these chemicals have been known to negatively affect humans who consume or breathe in pesticides. Accidental inhalation or consumption of a contaminated fish or animal tissue are the most likely ways to ingest organochloride pesticides. To confirm that someone is poisoned by a organochloride pesticide, blood or urine is typically sent to a university or government facility that uses GC/MS to test for the chemical compounds.

Warning signs of organochloride pesticide toxicity include seizures, hallucinations, coughing, skin rash, vomiting, abdominal pain, headaches, confusion and possibly respiratory failure, according to Matthew Wong, M.D., M.P.H., Beth Israel Deaconess Medical Center, on Medscape. Although there are bans in the U.S. and Europe on many of these pesticides, their use in other parts of the world and storage in parts of the U.S. and Europe create situations where poisonings are still possible.

Organochloride pesticides encompass a large number of durable chemicals that are both effective and carry a fair amount of risk across the globe.

Physical and Chemical Properties

Chlorination modifies the physical properties of hydrocarbons in several ways. The compounds are typically denser than water due to the higher atomic weight of chlorine versus hydrogen. Aliphatic organochlorides are alkylating agents because chloride is a leaving group.

Natural Occurrence

Many organochlorine compounds have been isolated from natural sources ranging from bacteria to humans. Chlorinated organic compounds are found in nearly every class of biomolecules including alkaloids, terpenes, amino acids, flavonoids, steroids, and fatty acids. Organochlorides, including dioxins, are produced in the high temperature environment of forest fires, and dioxins have been found in the preserved ashes of lightning-ignited fires that predate synthetic dioxins. In addition, a variety of simple chlorinated hydrocarbons including dichloromethane, chloroform, and carbon tetrachloride have been isolated from marine algae. A majority of the chloromethane in the environment is produced naturally by biological decomposition, forest fires, and volcanoes.

The natural organochloride epibatidine, an alkaloid isolated from tree frogs, has potent analgesic effects and has stimulated research into new pain medication. (The frogs obtain epibatidine through their diet and then sequester it on their skin. Likely dietary sources are beetles, ants, mites, and flies.)

Preparation

From Chlorine

Alkanes and aryl alkanes may be chlorinated under free radical conditions, with UV light. However, the extent of chlorination is difficult to control. Aryl chlorides may be prepared by the Friedel-Crafts halogenation, using chlorine and a Lewis acid catalyst.

The haloform reaction, using chlorine and sodium hydroxide, is also able to generate alkyl halides from methyl ketones, and related compounds. Chloroform was formerly produced thus.

Chlorine adds to the multiple bonds on alkenes and alkynes as well, giving di- or tetra-chloro compounds.

Reaction with Hydrogen Chloride

Alkenes react with hydrogen chloride (HCl) to give alkyl chlorides. For example, the industrial production of chloroethane proceeds by the reaction of ethylene with HCl:

$$H_2C = CH_2 + HCl \rightarrow CH_3CH_2Cl$$

In oxychlorination, hydrogen chloride instead of the more expensive chlorine for the same purpose:

$$CH_2 = CH_2 + 2HCl + \frac{1}{2}O_2 \rightarrow ClCH_2CH_2Cl + H_2O$$

Secondary and tertiary alcohols react with hydrogen chloride to give the corresponding chlorides. In the laboratory, the related reaction involving zinc chloride in concentrated hydrochloric acid:

$$\text{R-OH} + \text{HCl} \underset{\Delta}{\overset{ZnCl_2}{\rightarrow}} \overset{\text{alkyl}}{\text{R}} - \overset{\text{halide}}{\text{Cl}} + H_2O$$

Called the Lucas reagent, this mixture was once used in qualitative organic analysis for classifying alcohols.

Other Chlorinating Agents

Alkyl chlorides are most easily prepared by treating alcohols with thionyl chloride ($SOCl_2$) or phosphorus pentachloride (PCl_5), but also commonly with sulfuryl chloride (SO_2Cl_2) and phosphorus trichloride (PCl_3):

$$ROH + SOCl_2 \rightarrow RCl + SO_2 + HCl$$
$$3\,ROH + PCl_3 \rightarrow 3\,RCl + H_3PO_3$$
$$ROH + PCl_5 \rightarrow RCl + POCl_3 + HCl$$

In the laboratory, thionyl chloride is especially convenient, because the byproducts are gaseous. Alternatively, the Appel reaction can be used:

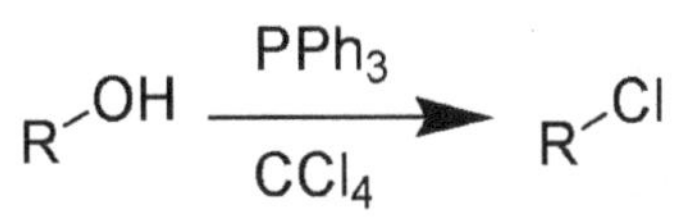

Reactions

Alkyl chlorides are versatile building blocks in organic chemistry. While alkyl bromides and iodides are more reactive, alkyl chlorides tend to be less expensive and more readily available. Alkyl chlorides readily undergo attack by nucleophiles.

Heating alkyl halides with sodium hydroxide or water gives alcohols. Reaction with alkoxides or aroxides give ethers in the Williamson ether synthesis; reaction with thiols give thioethers. Alkyl chlorides readily react with amines to give substituted amines. Alkyl chlorides are substituted by softer halides such as the iodide in the Finkelstein reaction. Reaction with other pseudohalides such as azide, cyanide, and thiocyanate are possible as well. In the presence of a strong base, alkyl chlorides undergo dehydrohalogenation to give alkenes or alkynes.

Alkyl chlorides react with magnesium to give Grignard reagents, transforming an electrophilic compound into a nucleophilic compound. The Wurtz reaction reductively couples two alkyl halides to couple with sodium.

Applications

Vinyl Chloride

The largest application of organochlorine chemistry is the production of vinyl chloride. The annual production in 1985 was around 13 billion kilograms, almost all of which was converted into polyvinylchloride (PVC).

Chloromethanes

Most low molecular weight chlorinated hydrocarbons such as chloroform, dichloromethane, dichloroethene, and trichloroethane are useful solvents. These solvents tend to be relatively non-polar; they are therefore immiscible with water and effective in cleaning applications such as degreasing and dry cleaning. Several billion kilograms of chlorinated methanes are produced annually, mainly by chlorination of methane:

$$CH_4 + XCl_2 \rightarrow CH_{4-X}\,Cl_X + XHCl$$

The most important is dichloromethane, which is mainly used as a solvent. Chloromethane is a precursor to chlorosilanes and silicones. Historically significant, but smaller in scale is chloroform, mainly a precursor to chlorodifluoromethane ($CHClF_2$) and tetrafluoroethene which is used in the manufacture of Teflon.

Pesticides

The two main groups of organochlorine insecticides are the DDT-type compounds and the chlorinated alicyclics. Their mechanism of action differs slightly.

- The DDT like compounds work on the peripheral nervous system. At the axon's sodium channel, they prevent gate closure after activation and membrane depolarization. Sodium ions leak through the nerve membrane and create a destabilizing negative "afterpotential" with hyperexcitability of the nerve. This leakage causes repeated discharges in the neuron either spontaneously or after a single stimulus.

- Chlorinated cyclodienes include aldrin, dieldrin, endrin, heptachlor, chlordane and endosulfan. A 2- to 8-hour exposure leads to depressed central nervous system (CNS) activity, followed by hyperexcitability, tremors, and then seizures. The mechanism of action is the insecticide binding at the $GABA_A$ site in the gamma-Aminobutyric acid (GABA) chloride ionophore complex, which inhibits chloride flow into the nerve.

- Other examples include dicofol, mirex, kepone, and pentachlorophenol. These can be either hydrophilic or hydrophobic, depending on their molecular structure.

Insulators

Polychlorinated biphenyls (PCBs) were once commonly used electrical insulators and heat transfer agents. Their use has generally been phased out due to health concerns. PCBs were replaced by polybrominated diphenyl ethers (PBDEs), which bring similar toxicity and bioaccumulation concerns.

Toxicity

Some types of organochlorides have significant toxicity to plants or animals, including humans. Dioxins, produced when organic matter is burned in the presence of chlorine, are persistent organic pollutantswhich pose dangers when they are released into the environment, as are some insecticides (such as DDT). For example, DDT, which was widely used to control insects in the mid-20th century, also accumulates in food chains, as do its metabolites DDE and DDD, and causes reproductive problems (e.g., eggshell thinning) in certain bird species. Some organochlorine compounds, such as sulfur mustards, nitrogen mustards, and Lewisite, are even used as chemical weapons due to their toxicity.

However, the presence of chlorine in an organic compound does not *ensure* toxicity. Some organochlorides are considered safe enough for consumption in foods and medicines. For example, peas and broad beans contain the natural chlorinated plant hormone 4-chloroindole-3-acetic acid (4-Cl-IAA); and the sweetener sucralose (Splenda) is widely used in diet products. As of 2004, at least 165 organochlorides had been approved worldwide for use as pharmaceutical drugs, including the natural antibiotic vancomycin, the antihistamine loratadine (Claritin), the antidepressant sertraline (Zoloft), the anti-epileptic lamotrigine (Lamictal), and the inhalation anesthetic isoflurane.

Rachel Carson brought the issue of DDT pesticide toxicity to public awareness with her 1962 book *Silent Spring*. While many countries have phased out the use of some types of organochlorides such as the US ban on DDT, persistent DDT, PCBs, and other organochloride residues continue to be found in humans and mammals across the planet many years after production and use have been limited. In Arcticareas, particularly high levels are found in marine mammals. These chemicals concentrate in mammals, and are even found in human breast milk. In some species of marine mammals, particularly those that produce milk with a high fat content, males typically have far higher levels, as females reduce their concentration by transfer to their offspring through lactation.

Exposure to Organochlorine Pesticides

People can be exposed to organochlorine pesticides through accidental inhalation exposure if you are in an area where they were recently applied. The chemicals can also be ingested in fish, dairy products, and other fatty foods that are contaminated. Organochlorine pesticides accumulate in the environment. They are very persistent and move long distances in surface runoff or groundwater. Prior to the mid-1970s, organochlorines resulted in widespread reproductive failure among birds because birds laid eggs with thin shells that cracked before hatching.

Effects of Pesticides on Humans

Exposure to organochlorine pesticides over a short period may produce convulsions, headache, dizziness, nausea, vomiting, tremors, confusion, muscle weakness, slurred speech, salivation and sweating. Long-term exposure to organochlorine pesticides may damage the liver, kidney, central nervous system, thyroid and bladder. Many of these pesticides have been linked to elevated rates of liver or kidney cancer in animals. There is some evidence indicating that organochlorine pesticides may also cause cancer in humans.

DDT

DDT, abbreviation of dichlorodiphenyltrichloroethane, also called 1,1,1-trichloro-2,2-bis (*p*-chlorophenyl)ethane, a synthetic insecticide belonging to the family of organic halogen compounds, highly toxic toward a wide variety of insects as a contact poison that apparently exerts its effect by disorganizing the nervous system.

DDT, prepared by the reaction of chloral with chlorobenzene in the presence of sulfuric acid, was first made in 1874; its insecticidal properties were discovered in 1939 by a Swiss chemist, Paul Hermann Müller. During and after World War II, DDT was found to be effective against lice, fleas, and mosquitoes (the carriers of typhus, of plague, and of malaria and yellow fever, respectively) as well as the Colorado potato beetle, the gypsy moth, and other insects that attack valuable crops.

Many species of insects rapidly develop populations resistant to DDT; the high stability of the compound leads to its accumulation in insects that constitute the diet of other animals, with toxic effects on them, especially certain birds and fishes. These two disadvantages had severely decreased the value of DDT as an insecticide by the 1960s, and severe restrictions were imposed on its use in the United States in 1972.

Pure DDT is a colourless, crystalline solid that melts at 109° C (228° F); the commercial product, which is usually 65 to 80 percent active compound, along with related substances, is an amorphous powder that has a lower melting point. DDT is applied as a dust or by spraying its aqueous suspension.

Properties and Chemistry

DDT is similar in structure to the insecticide methoxychlor and the acaricide dicofol. It is highly hydrophobic and nearly insoluble in water but has good solubility in most organic solvents, fats and oils. DDT does not occur naturally and is synthesised by a Friedel–Crafts hydroxyalkylation reaction between chloral (CCl_3CHO) and chlorobenzene (C_6H_5Cl), in the presence of an acidic catalyst. DDT has been marketed under trade names including Anofex, Cezarex, Chlorophenothane, Clofenotane, Dicophane, Dinocide, Gesarol, Guesapon, Guesarol, Gyron, Ixodex, Neocid, Neocidol and Zerdane.

Isomers and Related Compounds

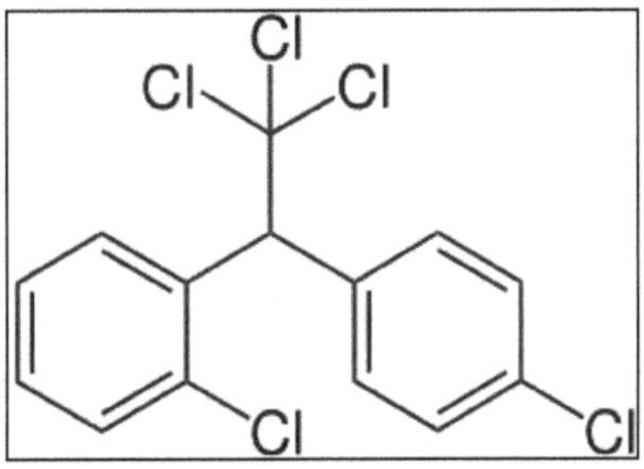

o,p'-DDT, a minor component in commercial DDT.

Commercial DDT is a mixture of several closely–related compounds. The major component (77%) is the *p,p'* isomer (pictured above). The *o,p'* isomer (pictured to the right) is also present in significant amounts (15%). Dichlorodiphenyldichloroethylene (DDE) and dichlorodiphenyldichloroethane (DDD) make up the balance. DDE and DDD are the major metabolites and environmental breakdown products.

Production and use

DDT has been formulated in multiple forms, including solutions in xylene or petroleum distillates, emulsifiable concentrates, water-wettable powders, granules, aerosols, smoke candles and charges for vaporizers and lotions.

From 1950 to 1980, DDT was extensively used in agriculture – more than 40,000 tonnes each year worldwide – and it has been estimated that a total of 1.8 million tonnes have been produced globally since the 1940s. In the United States, it was manufactured by some 15 companies, including Monsanto, Ciba,Montrose Chemical Company, Pennwalt, and Velsicol Chemical Corporation. Production peaked in 1963 at 82,000 tonnes per year. More than 600,000 tonnes (1.35 billion pounds) were applied in the US before the 1972 ban. Usage peaked in 1959 at about 36,000 tonnes.

In 2009, 3,314 tonnes were produced for malaria control and visceral leishmaniasis. India is the only country still manufacturing DDT, and is the largest consumer. China ceased production in 2007.

Mechanism of Insecticide Action

In insects, DDT opens sodium ion channels in neurons, causing them to fire spontaneously, which leads to spasms and eventual death. Insects with certain mutations in their sodium channel gene are resistant to DDT and similar insecticides. DDT resistance is also conferred by up-regulation of genes expressing cytochrome P450 in some insect species, as greater quantities of some enzymes of this group accelerate the toxin›s metabolism into inactive metabolites. (The same enzyme family is up-regulated in mammals too, e.g., in response to ethanol consumption.) Genomic studies in the model genetic organism *Drosophila melanogaster* revealed that high level DDT resistance is polygenic, involving multiple resistance mechanisms.

Environmental Impact

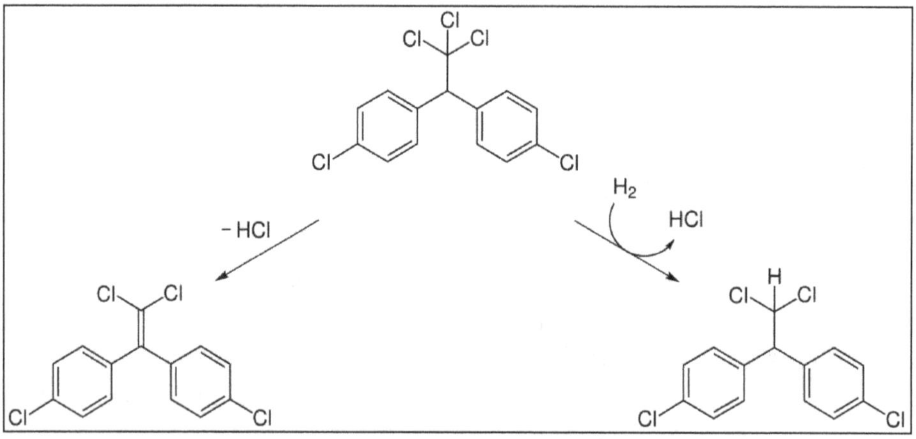

Degradation of DDT to form DDE (by elimination of HCl, left) and DDD (by reductive dechlorination, right).

DDT is a persistent organic pollutant that is readily adsorbed to soils and sediments, which can act both as sinks and as long-term sources of exposure affecting organisms. Depending on conditions, its soil half-life can range from 22 days to 30 years. Routes of loss and degradation include runoff, volatilization, photolysis and aerobic and anaerobic biodegradation. Due to hydrophobic properties, in aquatic ecosystems DDT and its metabolites are absorbed by aquatic organisms and adsorbed on suspended particles, leaving little DDT dissolved in the water (however, its half-life in aquatic environments is listed by the National Pesticide Information Center as 150 years). Its breakdown products and metabolites, DDE and DDD, are also persistent and have similar chemical and physical properties. DDT and its breakdown products are transported from warmer areas to the Arctic by the phenomenon of global distillation, where they then accumulate in the region's food web.

Medical researchers in 1974 found a measurable and significant difference in the presence of DDT in human milk between mothers who lived in New Brunswick and mothers who lived in Nova Scotia, "possibly because of the wider use of insecticide sprays in the past."

Because of its lipophilic properties, DDT can bioaccumulate, especially in predatory birds. DDT is toxic to a wide range of living organisms, including marine animals such as crayfish, daphnids, sea shrimp and many species of fish. DDT, DDE and DDD magnify through the food chain, with apex predators such as raptor birds concentrating more chemicals than other animals in the same environment. They are stored mainly in body fat. DDT and DDE are resistant to metabolism; in humans, their half-lives are 6 and up to 10 years, respectively. In the United States, these chemicals were detected in almost all human blood samples tested by the Centers for Disease Controlin 2005, though their levels have sharply declined since most uses were banned. Estimated dietary intake has declined, although FDA food tests commonly detect it.

Eggshell Thinning

The chemical and its breakdown products DDE and DDD caused eggshell thinning and population declines in multiple North American and European bird of prey species. DDE-related eggshell thinning is considered a major reason for the decline of the bald eagle, brown pelican, peregrine falcon and osprey. However, birds vary in their sensitivity to these chemicals, with birds of prey, waterfowl and song birds being more susceptible than chickens and related species. Even in 2010, California condors that feed on sea lions at Big Sur that in turn feed in the Palos Verdes Shelf area of the Montrose Chemical Superfund site exhibited continued thin-shell problems, though DDT›s role in the decline of the California condor is disputed.

The biological thinning mechanism is not entirely understood, but DDE appears to be more potent than DDT, and strong evidence indicates that p,p›-DDE inhibits calcium ATPase in the membrane of the shell gland and reduces the transport of calcium carbonate from blood into the eggshell gland. This results in a dose-dependent thickness reduction. Other evidence indicates that o,p›-DDT disrupts female reproductive tract development, later impairing eggshell quality. Multiple mechanisms may be at work, or different mechanisms may operate in different species.

Human Health

DDT is an endocrine disruptor. It is considered likely to be a human carcinogen although the majority of studies suggest it is not directly genotoxic. DDEacts as a weak androgen receptor antagonist, but not as an estrogen. p,p'-DDT, DDT's main component, has little or no androgenic or estrogenic activity. The minor component o,p'-DDT has weak estrogenic activity.

A U.S. soldier is demonstrating DDT hand-spraying equipment. DDT was used to control the spread of typhus-carrying lice.

Acute Toxicity

DDT is classified as "moderately toxic" by the US National Toxicology Program (NTP) and "moderately hazardous" by WHO, based on the rat oral LD_{50} of 113 mg/kg. Indirect exposure is considered relatively non-toxic for humans.

Chronic Toxicity

Primarily through the tendency for DDT to buildup in areas of the body with high lipid content, chronic exposure can affect reproductive capabilities and the embryo or fetus.

- Research has shown that exposure to DDT at amounts that would be needed in malaria control might cause preterm birth and early weaning; toxicological evidence shows endocrine-disrupting properties; human data also indicate possible disruption in semen quality, menstruation, gestational length, and duration of lactation.

- Other studies document decreases in semen quality among men with high exposures (generally from IRS).

- Studies are inconsistent on whether high blood DDT or DDE levels increase time to pregnancy. In mothers with high DDE blood serum levels, daughters may have up to a 32% increase in the probability of conceiving, but increased DDT levels have been associated with a 16% decrease in one study.

- Indirect exposure of mothers through workers directly in contact with DDT is associated with an increase in spontaneous abortions.

- Other studies found that DDT or DDE interfere with proper thyroid function in pregnancy and childhood.

Carcinogenicity

In 2015, the International Agency for Research on Cancer classified DDT as Group 2A "probably carcinogenic to humans". Previous assessments by the U.S. National Toxicology Program classified it as "reasonably anticipated to be a carcinogen" and by the EPA classified DDT, DDE and DDD as class B2 "probable" carcinogens; these evaluations were based mainly on animal studies.

Occupational DDT exposure was associated with increased pancreatic cancer risk in 2 case control studies, but another study showed no DDE dose-effect association. Results regarding a possible association with liver cancer and biliary tract cancer are conflicting: workers who did not have direct occupational DDT contact showed increased risk. White men had an increased risk, but not white women or black men. Results about an association with multiple myeloma, prostate and testicular cancer, endometrial cancer and colorectal cancer have been inconclusive or generally do not support an association. A 2017 review of liver cancer studies concluded that "organochlorine pesticides, including DDT, may increase hepatocellular carcinoma risk."

A 2009 review, whose co-authors included persons engaged in DDT-related litigation, reached broadly similar conclusions, with an equivocal association with testicular cancer. Case–control studies did not support an association with leukemia or lymphoma.

Breast Cancer

The question of whether DDT or DDE are risk factors in breast cancer has not been conclusively answered. Several meta analyses of observational studies have concluded that there is no overall relationship between DDT exposure and breast cancer risk. The United States Institute of Medicine reviewed data on the association of breast cancer with DDT exposure in 2012 and concluded that a causative relationship could neither be proven nor disproven.

A 2007 case control study using archived blood samples found that breast cancer risk was increased 5-fold among women who were born prior to 1931 and who had high serum DDT levels in 1963. Reasoning that DDT use became widespread in 1945 and peaked around 1950, they concluded that the ages of 14–20 were a critical period in which DDT exposure leads to increased risk. This study, which suggests a connection between DDT exposure and breast cancer that would not be picked up by most studies, has received variable commentary in third party reviews. One review suggested that "previous studies that measured exposure in older women may have missed the critical period." The National Toxicology Program notes that while the majority of studies have not found a relationship between DDT exposure and breast cancer that positive associations have been seen in a "few studies among women with higher levels of exposure and among certain subgroups of women".

A 2015 case control study identified a link (odds ratio 3.4) between *in-utero* exposure (as estimated from archived maternal blood samples) and breast cancer diagnosis in daughters. The findings "support classification of DDT as an endocrine disruptor, a predictor of breast cancer, and a marker of high risk".

Malaria Control

Malaria remains the primary public health challenge in many countries. In 2015, there were 214 million cases of malaria worldwide resulting in an estimated 438,000 deaths, 90% of which occurred in Africa.DDT is one of many tools to fight the disease. Its use in this context has been called everything from a "miracle weapon [that is] like Kryptonite to the mosquitoes," to "toxic colonialism".

Before DDT, eliminating mosquito breeding grounds by drainage or poisoning with Paris green or pyrethrum was sometimes successful. In parts of the world with rising living standards, the elimination of malaria was often a collateral benefit of the introduction of window screens and improved sanitation. A variety of usually simultaneous interventions represents best practice. These include antimalarial drugs to prevent or treat infection; improvements in public health infrastructure to diagnose, sequester and treat infected individuals; bednets and other methods intended to keep mosquitoes from biting humans; and vector control strategies such as larvaciding with insecticides, ecological controls such as draining mosquito breeding grounds or introducing fish to eat larvae and indoor residual spraying (IRS) with insecticides, possibly including DDT. IRS involves the treatment of interior walls and ceilings with insecticides. It is particularly effective against mosquitoes, since many species rest on an indoor wall before or after feeding. DDT is one of 12 WHO–approved IRS insecticides.

WHO's anti-malaria campaign of the 1950s and 1960s relied heavily on DDT and the results were promising, though temporary in developing countries. Experts tie malarial resurgence to multiple factors, including poor leadership, management and funding of malaria control programs; poverty; civil unrest; and increased irrigation. The evolution of resistance to first-generation drugs (e.g. chloroquine) and to insecticides exacerbated the situation. Resistance was largely fueled by unrestricted agricultural use. Resistance and the harm both to humans and the environment led many governments to curtail DDT use in vector control and agriculture. In 2006 WHO reversed a long-standing policy against DDT by recommending that it be used as an indoor pesticide in regions where malaria is a major problem.

Once the mainstay of anti-malaria campaigns, as of 2008 only 12 countries used DDT, including India and some southern African states, though the number was expected to rise.

Initial Effectiveness

When it was introduced in World War II, DDT was effective in reducing malaria morbidity and mortality. WHO's anti-malaria campaign, which consisted mostly of spraying DDT and rapid treatment and diagnosis to break the transmission cycle, was initially successful as well. For example, in Sri Lanka, the program reduced cases from about one million per year before spraying to just 18 in 1963 and 29 in 1964. Thereafter the program was halted to save money and malaria rebounded to 600,000 cases in 1968 and the first quarter of 1969. The country resumed DDT vector control but the mosquitoes had evolved resistance in the interim, presumably because of continued agricultural use. The program switched to malathion, but despite initial successes, malaria continued its resurgence into the 1980s.

DDT remains on WHO's list of insecticides recommended for IRS. After the appointment of Arata Kochi as head of its anti-malaria division, WHO's policy shifted from recommending IRS only in areas of seasonal or episodic transmission of malaria, to advocating it in areas of continuous, intense transmission. WHO reaffirmed its commitment to phasing out DDT, aiming "to achieve a 30% cut in the application of DDT world-wide by 2014 and its total phase-out by the early 2020s if not sooner" while simultaneously combating malaria. WHO plans to implement alternatives to DDT to achieve this goal.

South Africa continues to use DDT under WHO guidelines. In 1996, the country switched to alternative insecticides and malaria incidence increased dramatically. Returning to DDT and introducing new drugs brought malaria back under control. Malaria cases increased in South America after countries in that continent stopped using DDT. Research data showed a strong negative relationship between DDT residual house sprayings and malaria. In a research from 1993 to 1995, Ecuador increased its use of DDT and achieved a 61% reduction in malaria rates, while each of the other countries that gradually decreased its DDT use had large increases.

Mosquito Resistance

In some areas resistance reduced DDT's effectiveness. WHO guidelines require that absence of resistance must be confirmed before using the chemical. Resistance is largely due to agricultural use, in much greater quantities than required for disease prevention.

Resistance was noted early in spray campaigns. Paul Russell, former head of the Allied Anti-Malaria campaign, observed in 1956 that "resistance has appeared after six or seven years." Resistance has been detected in Sri Lanka, Pakistan, Turkey and Central America and it has largely been replaced by organophosphate or carbamate insecticides, *e.g.* malathion or bendiocarb.

In many parts of India, DDT is ineffective. Agricultural uses were banned in 1989 and its anti-malarial use has been declining. Urban use ended. One study concluded that

"DDT is still a viable insecticide in indoor residual spraying owing to its effectivity in well supervised spray operation and high excito-repellency factor."

Studies of malaria-vector mosquitoes in KwaZulu-Natal Province, South Africa found susceptibility to 4% DDT (WHO›s susceptibility standard), in 63% of the samples, compared to the average of 87% in the same species caught in the open. The authors concluded that "Finding DDT resistance in the vector *An. arabiensis*, close to the area where we previously reported pyrethroid-resistance in the vector *An. funestus* Giles, indicates an urgent need to develop a strategy of insecticide resistance management for the malaria control programmes of southern Africa."

DDT can still be effective against resistant mosquitoes and the avoidance of DDT-sprayed walls by mosquitoes is an additional benefit of the chemical. For example, a 2007 study reported that resistant mosquitoes avoided treated huts. The researchers argued that DDT was the best pesticide for use in IRS (even though it did not afford the most protection from mosquitoes out of the three test chemicals) because the others pesticides worked primarily by killing or irritating mosquitoes – encouraging the development of resistance. Others argue that the avoidance behavior slows eradication. Unlike other insecticides such as pyrethroids, DDT requires long exposure to accumulate a lethal dose; however its irritant property shortens contact periods. "For these reasons, when comparisons have been made, better malaria control has generally been achieved with pyrethroids than with DDT." In India outdoor sleeping and night duties are common, implying that "the excito-repellent effect of DDT, often reported useful in other countries, actually promotes outdoor transmission."

Residents' Concerns

IRS is effective if at least 80% of homes and barns in a residential area are sprayed. Lower coverage rates can jeopardize program effectiveness. Many residents resist DDT spraying, objecting to the lingering smell, stains on walls, and the potential exacerbation of problems with other insect pests. Pyrethroid insecticides (e.g. deltamethrin and lambda-cyhalothrin) can overcome some of these issues, increasing participation.

Human Exposure

A 1994 study found that South Africans living in sprayed homes have levels that are several orders of magnitude greater than others. Breast milk from South African mothers contains high levels of DDT and DDE. It is unclear to what extent these levels arise from home spraying vs food residues. Evidence indicates that these levels are associated with infant neurological abnormalities.

Most studies of DDT's human health effects have been conducted in developed countries where DDT is not used and exposure is relatively low.

Illegal diversion to agriculture is also a concern as it is difficult to prevent and its subsequent use on crops is uncontrolled. For example, DDT use is widespread in Indian agriculture, particularly mangoproduction and is reportedly used by librarians to protect books. Other examples include Ethiopia, where DDT intended for malaria control is reportedly used in coffee production, and Ghana where it is used for fishing." The residues in crops at levels unacceptable for export have been an important factor in bans in several tropical countries. Adding to this problem is a lack of skilled personnel and management.

Criticism of Restrictions on DDT use

A few people and groups have argued that limitations on DDT use for public health purposes have caused unnecessary morbidity and mortality from vector-borne diseases, with some claims of malaria deaths ranging as high as the hundreds of thousands and millions. Robert Gwadz of the US National Institutes of Health said in 2007, "The ban on DDT may have killed 20 million children." These arguments were rejected as "outrageous" by former WHO scientist Socrates Litsios. May Berenbaum, University of Illinois entomologist, says, "to blame environmentalists who oppose DDT for more deaths than Hitler is worse than irresponsible."

Criticisms of a DDT "ban" often specifically reference the 1972 United States ban (with the erroneous implication that this constituted a worldwide ban and prohibited use of DDT in vector control). Reference is often made to *Silent Spring,* even though Carson never pushed for a DDT ban. John Quiggin and Tim Lambert wrote, "the most striking feature of the claim against Carson is the ease with which it can be refuted."

Alternatives

Insecticides

Organophosphate and carbamate insecticides, *e.g.* malathion and bendiocarb, respectively, are more expensive than DDT per kilogram and are applied at roughly the same dosage. Pyrethroids such as deltamethrin are also more expensive than DDT, but are applied more sparingly (0.02–0.3 g/m^2 vs 1–2 g/m^2), so the net cost per house is about the same.

Non-chemical Vector Control

Before DDT, malaria was successfully eliminated or curtailed in several tropical areas by removing or poisoning mosquito breeding grounds and larva habitats, for example by eliminating standing water. These methods have seen little application in Africa for more than half a century. According to CDC, such methods are not practical in Africa because "*Anopheles gambiae*, one of the primary vectors of malaria in Africa, breeds in numerous small pools of water that form due to rainfall. It is difficult, if not impossible, to predict when and where the breeding sites will form, and to find and treat them before the adults emerge."

The relative effectiveness of IRS versus other malaria control techniques (e.g. bednets or prompt access to anti-malarial drugs) varies and is dependent on local conditions.

A WHO study released in January 2008 found that mass distribution of insecticide-treated mosquito nets and artemisinin–based drugs cut malaria deaths in half in malaria-burdened Rwanda and Ethiopia. IRS with DDT did not play an important role in mortality reduction in these countries.

Vietnam has enjoyed declining malaria cases and a 97% mortality reduction after switching in 1991 from a poorly funded DDT-based campaign to a program based on prompt treatment, bednets and pyrethroid group insecticides.

In Mexico, effective and affordable chemical and non-chemical strategies were so successful that the Mexican DDT manufacturing plant ceased production due to lack of demand.

A review of fourteen studies in sub-Saharan Africa, covering insecticide-treated nets, residual spraying, chemoprophylaxis for children, chemoprophylaxis or intermittent treatment for pregnant women, a hypothetical vaccine and changing front–line drug treatment, found decision making limited by the lack of information on the costs and effects of many interventions, the small number of cost-effectiveness analyses, the lack of evidence on the costs and effects of packages of measures and the problems in generalizing or comparing studies that relate to specific settings and use different methodologies and outcome measures. The two cost-effectiveness estimates of DDT residual spraying examined were not found to provide an accurate estimate of the cost-effectiveness of DDT spraying; the resulting estimates may not be good predictors of cost-effectiveness in current programs.

However, a study in Thailand found the cost per malaria case prevented of DDT spraying (US$1.87) to be 21% greater than the cost per case prevented of lambda-cyhalothrin–treated nets (US$1.54),casting some doubt on the assumption that DDT was the most cost-effective measure. The director of Mexico's malaria control program found similar results, declaring that it was 25% cheaper for Mexico to spray a house with synthetic pyrethroids than with DDT. However, another study in South Africa found generally lower costs for DDT spraying than for impregnated nets.

A more comprehensive approach to measuring cost-effectiveness or efficacy of malarial control would not only measure the cost in dollars, as well as the number of people saved, but would also consider ecological damage and negative human health impacts. One preliminary study found that it is likely that the detriment to human health approaches or exceeds the beneficial reductions in malarial cases, except perhaps in epidemics. It is similar to the earlier study regarding estimated theoretical infant mortality caused by DDT.

A study in the Solomon Islands found that "although impregnated bed nets cannot entirely replace DDT spraying without substantial increase in incidence, their use permits reduced DDT spraying."

A comparison of four successful programs against malaria in Brazil, India, Eritrea and Vietnam does not endorse any single strategy but instead states, "Common success factors included conducive country conditions, a targeted technical approach using a package of effective tools, data-driven decision-making, active leadership at all levels of government, involvement of communities, decentralized implementation and control of finances, skilled technical and managerial capacity at national and sub-national levels, hands-on technical and programmatic support from partner agencies, and sufficient and flexible financing."

DDT resistant mosquitoes have generally proved susceptible to pyrethroids. Thus far, pyrethroid resistance in *Anopheles* has not been a major problem.

DDT Accumulation

Fat Stores:

- DDT tends to accumulate in the fatty tissues of insects, wildlife, and people, but produces no known toxic effects while it is stored in the fat.
- DDT is metabolized into various breakdown products in the body including DDE, DDD[4] , and DDA[5].
- When fat stores are used during periods of starvation the breakdown products of DDT are released into the blood where they may be toxic to the liver and the nervous system.
- Once DDT has accumulated in the body, it is excreted in the urine, feces, or breast milk. Breast milk is often used to measure a population's exposure to DDT.

Biomagnification:

- Because of DDT's chemical properties it has the tendency to accumulate in animals. As animals lower on the food chain are eaten by other animals higher up, DDT becomes concentrated in the fatty tissues of predators. This continues until the primary predator of the food chain receives the highest dose, which may lead to adverse health effects. Once the use of DDT was discontinued in the U.S., its concentration in the environment and animals decreased.

Aldrin

Aldrin is a pesticide used to control soil insects such as termites, corn rootworm, wireworms, rice water weevil, and grasshoppers. It has been widely used to protect crops such as corn and potatoes, and has been effective to protect wooden structures from termites.

Aldrin is readily metabolized to dieldrin by both plants and animals. As a result, Aldrin residues are rarely found in foods and animals, and then only in small amounts. It binds strongly to soil particles and is very resistant to leaching into groundwater. Volatilization is an important mechanism of loss from soils. Due to its persistent nature and hydrophobicity, Aldrin is known to bioconcentrate, mainly as its conversion products.

Aldrin is banned in many countries, including Bulgaria, Ecuador, Finland, Hungary, Israel, Singapore, Switzerland, and Turkey. Its use is severely restricted in many countries, including Argentina, Austria, Canada, Chile, the EU, Japan, New Zealand, the Philippines, USA, and Venezuela.

Use in South East Asia

	Used or Found in Country?	Years of Usage	Regulatory Controls
Cambodia			Banned in 1992
Lao PDR			Banned in 1992
Malaysia	Yes	1980-1994	Banned in 1994
Thailand	Yes	1955	Banned in 1988
Indonesia			Never registered for use
Philippines			Banned in 1989
Viet Nam			Banned in 1992

Effects on Plants and Animals

Aldrin has low phytotoxicity with plants affected only by extremely high application rates. The toxicity of Aldrin to aquatic organisms is quite variable, with aquatic insects being the most sensitive group of invertebrates. The 96-h LC50 dose values range from 1-200 μg/L for insects, and from 2.2-53 μg/L for fish. Long term and bioconcentration studies are performed primarily using Dieldrin, the primary conversion product of Aldrin. In a model ecosystem study, only 0.5% of the original radioactive Aldrin was stored as Aldrin in the mosquitofish (Gambusia affinis), the organism at the top of the model food chain.

The acute toxicity of Aldrin to avian species varies in the range of 6.6 mg/kg for bobwhite quail to 520 mg/kg for mallard ducks. Aldrin treated rice is thought to have been the cause of deaths in waterfowl, shorebirds and passerines along the Texas Gulf Coast, both by direct poisoning by ingestion of Aldrin treated rice and indirectly by consuming organisms contaminated with Aldrin. Residues of Aldrin were detected in all samples of bird casualties, eggs, scavengers, predators, fish, frogs, invertebrates and soil.

As Aldrin is readily and rapidly converted to Dieldrin in the environment its fate is closely linked to that of Dieldrin. Aldrin is readily metabolised to dieldrin in both animals and plants, and therefore Aldrin residues are rarely present in animals and then only in very small amounts. Residues of Aldrin have been detected in fish in Egypt, the average concentration was 8.8 μg/kg, and a maximum concentration of 54.27 μg/kg.

Effects on Humans

Aldrin is toxic to humans; the lethal dose of Aldrin for an adult man has been estimated to be about 5g, equivalent to 83 mg/kg body weight. Signs and symptoms of Aldrin intoxication may include headache, dizziness, nausea, general malaise and vomiting, followed by muscle twitching, myoclonic jerks, and convulsions. Occupational exposure to Aldrin, in conjunction with Dieldrin and Endrin, was associated with a significant increase in liver and biliary cancer, although the study did have some limitations, including a lack of quantitative exposure information. There is limited information that cyclodienes, such as Aldrin, may affect immune responses.

The acute oral LD50 dose for Aldrin in laboratory animals is in the range of 33 mg/kg body weight for guinea pigs to 320 mg/kg body weight for hamsters. Reproductive effects in rats were observed when pregnant females were dosed with 1.0 mg/kg Aldrin subcutaneously. Offspring experienced a decrease in the median effective time for incisor teeth eruption and an increase in the median effective time for testes descent. There is, as yet, no evidence of a teratogenic potential for Aldrin. The International Agency for Research on Cancer (IARC) has concluded that there is inadequate evidence for the carcinogenicity of Aldrin in humans, and there is only limited evidence in experimental animals. Aldrin is therefore not classifiable as to its carcinogenicity in humans.

The average daily human intake of Aldrin and Dieldrin was calculated to be 19μg/person in India, and 0.55 μg/person in Vietnam. Dairy products, such as milk and butter, and animal meats are the primary sources of exposure.

Endrin

Endrin is a foliar insecticide used mainly on field crops such as cotton and grains. It has also been used as a rodenticide to control mice and voles. It is rapidly metabolised by animals and does not accumulate in fat to the same extent as other compounds with similar structures. It can enter the atmosphere by volatilization, and can contaminate surface water from soil run-off.

Endrin is banned in many countries, including Belgium, Cyprus, Ecuador, Finland, Israel, Philippines, Singapore, Thailand, and Togo. Its use is severely restricted in many countries, including Argentina, Canada, Chile, Colombia, the EU, India, Japan, New Zealand, Pakistan, USA, and Venezuela.

Usage in South East Asia

	Used or Found in Country?	Years of Usage	Regulatory Controls
Cambodia	Yes	1950-92 Present?	Banned
Lao PDR			Banned in 1992
Malaysia			Withdrawn by company
Thailand	Yes	1955	Banned in 1981
Indonesia			Never registered for use
Philippines			Banned from import in 1989
Viet Nam			Banned in 1992

Potential Effects on Humans

A study of workers involved in the production of Aldrin, Dieldrin and Endrin did not find Endrin in the blood of workers, except in cases of accidental, acute over-exposure. These findings are in agreement with results of a study of 71 workers in an Endrin plant in the USA. Data on absenteeism, results of liver function tests, blood chemistry, blood morphology, urine analysis, occurrence of sensitization, and the incidence and pattern of diseases, including the occurrence of malignant growth, showed no difference between workers exposed to Endrin and other chemical plant operators. A study of workers involved in the manufacture of Aldrin, Dieldrin and Endrin found a statistically significant increase in liver and biliary tract cancers, although the study did have some limitations such as a lack of quantitative exposure information. There is limited evidence that cyclodienes such as Endrin may also depress immune responses.

IARC has concluded that there is inadequate evidence for the carcinogenicity of Endrin in humans, and there is only limited evidence in experimental animals. Endrin is therefore not classifiable as to its carcinogenicity in humans.

Potential Effects on Animals

Endrin is metabolised rapidly by animals, and very little is accumulated in fat compared to compounds of similar structure (including its stereoisomer - Dieldrin). The formation of anti-12-hydroxyEndrin is considered to be the major route of metabolism of Endrin.

The acute oral LD50 of Endrin is in the range of 3 mg/kg body weight in monkeys to 36 mg/kg in guinea pigs.

Male and female Long-Evans rats were fed Endrin in their diet over three generations. No difference in appearance, behaviour, body weight, or number or size of litters was observed. The weights of liver, kidneys and brain were normal, and no histopathological abnormalities were observed in third generation weanlings.

Significant increased mortality of pups in the second and third generations of rats fed 3 mg/kg was noted. Endrin was not teratogenic at levels that did not cause maternal toxicity.

Endrin is highly toxic to fish, with most LC50 values below 1.0 μg/L. Sheepshead minnow embryos exposed for 23 weeks to 0.31 and 0.72 μg/L hatched early, and all those exposed to 0.72 μg/L died by the ninth day of their exposure, while those exposed at 0.31 μg/L were initially stunted and some died. The reproductive ability of the survivors of the 0.31 μg/L was impaired. No significant effects were observed at an exposure concentration of 0.12 μg/L. The lowest observed adverse effect level (LOAEL) for aquatic organisms was 30 ng/L over 20 days for reproduction in mysid shrimp. Reproduction in male and female mallard ducks was not impaired by diets containing 0, 0.5 or 3.0 mg/kg.

Environmental Behavior

Insecticides like dieldrin and endrin have been shown to persist for decades in the environment. A definitive detection of the residues was not possible until 1971 when mass spectrometer started being used as a detector in gas chromatography. Detection of these chemicals in the environment has been reported across the world up to 2005, even though the frequency of reported cases are low due to its relatively small-scale use and very low concentrations.

Endrin regularly enters the environment when applied to crops or when rain washes it off. It has been found in water, sediments, atmospheric air and biotic environment, even after uses have been stopped.Organochlorine pesticides strongly resist degradation, are poorly soluble in water but highly soluble in lipids, which is called lipophilic. This leads to bioaccumulation in fatty tissues of organisms, mainly those dwelling in water. A high bioconcentration factor of 1335-10,000 has been reported in fish. Endrin binds very strongly to organic matter in soil and aquatic sediments due to their high adsorption coefficient,making it less likely to leach into groundwater, even though contaminated groundwater samples have been found. In 2009, EPA released data indicating that the endrin in soil could last up to 14 years or more. The extent of endrin›s persistence depends highly on local conditions. For example, high temperature (230 °C) or intense sunlight leads to more rapid breakdown of endrin into endrin ketone and endrin aldehyde, however, this breakdown is less than 5%.

Removal from the Environment

In the United States, endrin was mainly disposed in land until U.S. federal regulations were applied in 1987 on land disposal of wastes containing endrin. Primary methods of endrin disappearance from soil are volatilization and photodecomposition. Under ultraviolet light, endrin forms δ-ketoendrin and International Programme on Chemical

Safety (IPCS) claims that in intense summer sun, about 50% of endrin is isomerized to δ-ketoendrin in 7 days. In anaerobic conditions microbial degradation by fungi and bacteria takes place to form the same major end product.

Hazardous Substances Data Bank (HSDB) lists reductive dechlorination and incineration for field disposal of small quantities of endrin. In reductive dechlorination, endrin›s chlorine atoms were completely replaced with hydrogen atoms, which is suspected to be more environmentally acceptable. Even though endrin binds very strongly to soil, phytoremediation has been proposed by group of Japanese scientists using crops in the Cucurbitaceae family. As of 2009, exact mechanisms behind the plant uptake of endrin have not been understood. Research in uptake mechanisms and factors that influence the uptake is needed for practical application.

Heptachlor

Heptachlor was used as an insecticide; however, nearly all registered uses of heptachlor have been canceled. Very limited information is available on the health effects of heptachlor in humans. Acute (short-term) inhalation exposure to heptachlor may result in nervous system effects, with oral studies showing gastrointestinal effects. Animal studies have reported effects on the liver and central nervous system from oral exposure. Chronic (long-term) inhalation and oral exposure by humans may be associated with neurological effects including irritability, salivation, and dizziness, while oral exposure may result in effects on the blood. Human studies are inconclusive regarding heptachlor and cancer. Animal oral studies have reported liver tumors. EPA has classified heptachlor as a Group B2, probable human carcinogen.

Uses

- Heptachlor is a constituent of technical grade chlordane, approximately 10 percent by weight.
- Heptachlor was used as an insecticide in the United States from 1953 to 1974. In 1974, nearly all registered uses of heptachlor were canceled.
- Heptachlor was used from 1953 to 1974 as a soil and seed treatment to protect corn, small grains, and sorghum from pests. It was also used to control ants, cutworms, maggots, termites, and other pests in agriculture and in the home.
- The sale of heptachlor was voluntarily canceled in 1987 by its sole U.S. manufacturer. In 1988, the sale, distribution, and shipment of existing stocks of all canceled heptachlor and chlordane products was prohibited in the United States.
- The only commercial use of heptachlor products still permitted is fire ant control in power transformers. In addition, homeowner's use of existing stocks of heptachlor-containing termite control products is also allowed.

Sources and Potential Exposure

- People whose homes were treated for termites with heptachlor may be exposed to heptachlor in the indoor air for many years after treatment. Workers who use heptachlor to kill fire ants or who manufacture the chemical may be exposed to it in the air or through the skin.
- Heptachlor has been detected in food, including fish, shellfish, dairy products, meat, and poultry. Another possible source of exposure is drinking water; heptachlor has been detected at low concentrations in drinking water wells in several states.

Assessing Personal Exposure

- Laboratory tests can detect heptachlor (and its breakdown product, heptachlor epoxide) in blood, fat, breast milk, and body tissues after exposure to high levels.

Health Hazard Information

Acute Effects:

- Acute inhalation exposure to heptachlor in humans has been associated with nervous system effects in a few case studies, while gastrointestinal effects, such as nausea and vomiting, have been reported to occur following accidental ingestion of heptachlor.
- Effects on the liver and central nervous system have been noted in animals acutely exposed to heptachlor via the oral route.
- Heptachlor is considered to have high to extreme acute toxicity based on short-term oral tests in rats.

Chronic Effects (Noncancer):

- Chronic inhalation exposure to heptachlor has been associated with blood effects in humans, while oral exposure has resulted in neurological effects including irritability, salivation, dizziness, muscle tremors, and convulsions.
- Animal studies have reported effects on the liver, kidney, and the immune and nervous systems from oral exposure to heptachlor.
- The Reference Dose (RfD) for heptachlor is 0.0005 milligrams per kilogram body weight per day (mg/kg/d) based on liver weight increases in males rats only. The RfD is an estimate (with uncertainty spanning perhaps an order of magnitude) of a daily oral exposure to the human population (including sensitive subgroups) that is likely to be without appreciable risk of deleterious

noncancer effects during a lifetime. It is not a direct estimator of risk but rather a reference point to gauge the potential effects. At exposures increasingly greater than the RfD, the potential for adverse health effects increases. Lifetime exposure above RfD does not imply that an adverse health effect would necessarily occur.

- EPA has low confidence in the study on which the RfD is based because it is of low quality; low confidence in the database because it is incomplete; and, consequently, low confidence in the RfD.
- EPA has not established a Reference Concentration (RfC) for heptachlor.

Reproductive/Developmental Effects:

- Heptachlor has been shown to cross the placenta to the developing fetus in humans. However, inadequate information is available to determine whether heptachlor may cause developmental or reproductive effects in humans.
- Animal studies have reported developmental effects, including fetal resorptions, and decreased postnatal survival, as well as reproductive effects such as failure of animals to reproduce, following oral exposure to heptachlor.

Cancer Risk:

- Human studies on heptachlor exposure and cancer are inconclusive. There are several case reports describing a possible link between heptachlor exposure and leukemia and neuroblastoma; however, insufficient information is available to confirm a causal effect. Several studies on workers exposed via inhalation to heptachlor are available; however, these are limited due to confounding factors and small sample size.
- Animal studies have reported liver tumors in mice exposed to heptachlor via ingestion.
- EPA considers heptachlor to be a probable human carcinogen (cancer-causing agent) and has classified it as a Group B2 carcinogen.
- EPA uses mathematical models, based on animal studies, to estimate the probability of a person developing cancer from breathing air containing a specified concentration of a chemical. EPA calculated an inhalation unit risk estimate of 1.3 $\times 10^{-3}$ ($\mu g/m^{3-1}$) . EPA estimates that, if an individual were to continuously breathe air containing heptachlor at an average of 0.0008 $\mu g/m^3$ (8 x 10^{-3} milligrams per cubic meter [mg/m^3]) over his or her entire lifetime, that person would theoretically have no more than a one-in-a-million increased chance of developing cancer as a direct result of breathing air containing this chemical. Similarly, EPA estimates that breathing air containing 0.008 $\mu g/m$ (8 x 10 mg/m^3) would result in not greater than 3 a one-in-a-hundred thousand increased chance of developing

cancer, and air containing 0.08 μg/m (8×10^{-5} mg/m^3) would result in not greater than a one-in-ten-thousand increased chance of developing cancer.

- EPA has calculated an oral cancer slope factor of 4.5 (mg/kg/d)$^{-1}$.

Organophosphate Insecticides

Organophosphates are a group of human-made chemicals that poison insects and mammals. Organophosphates are the most widely used insecticides today. They are used in agriculture, the home, gardens, and veterinary practice.

Organophosphate insecticides (such as diazinon) are one type of pesticide that works by damaging an enzyme in the body called acetylcholinesterase. This enzyme is critical for controlling nerve signals in the body. The damage to this enzyme kills pests and may cause unwanted side effects in exposed humans. All organophosphates have a common mechanism of toxicity and can cause similar symptoms in humans who have too much exposure.

Urine samples from participants in the Churchill County leukemia study were analyzed for organophosphate insecticides and their breakdown products (metabolites). The metabolites are not toxic, but show that exposure to organophosphates occurred in the few days before testing. Linking some of these metabolites to a specific, original organophosphate compound is not possible without additional information. Even though an organophosphate or its metabolites in a person indicates exposure, it does not necessarily show the person will become sick.

In a study of leukemia in children, levels of two organophosphate metabolites were high:

- 3,5,6-Trichloro-2-pyridinol, a metabolite of chlorpyrifos.
- Diethylthiophosphate, a metabolite of numerous organophosphates such as diazinon.

Effects of Organophosphates on Environment

- Because organophosphates are often sprayed on crops and plants, small particles of the chemical may be carried away from the field or yard before falling to the ground.
- After organophosphates are applied, they may be present in the soil, surface waters, and on the surface of the plants. They can move through the soil and contaminate ground water.
- Rain can wash organophosphates on soil and plant surfaces into surface waters.

- Organophosphates are rapidly broken down into other chemicals so they do not build up in the environment.
- Organophosphates are not likely to build up to high or dangerous levels in animal or plant foods that you might eat.

Ways of Getting Exposed to Organophosphates

- By working for companies that make or apply organophosphates.
- By ingesting (eating or drinking) or breathing them or by getting them in your eyes or on your skin.
- By touching contaminated soils or contaminated runoff water or groundwater.
- Eating contaminated food and skin contact during home application are the main sources of exposure.

Effects of Organophosphates on Humans

Symptoms of sudden poisoning by organophosphates start during or after exposure, depending on how the poison is contacted. Symptoms start fastest after organophosphates are breathed, and next by eating or drinking contaminated food or water or getting them on your skin. Some symptoms are headache, dizziness, weakness, diarrhea, nausea and vomiting, salivation, watery eyes, and small pupils. Severe symptoms are seizures, slow pulse, difficulty breathing, and coma. Long after exposure, people also can develop nervous system problems such as muscle weakness and numbness and tingling of the hands and feet (neuropathy).

Long-term exposure to organophosphates can cause confusion, anxiety, loss of memory, loss of appetite, disorientation, depression, and personality changes. Other symptoms such as weakness, headache, diarrhea, nausea and vomiting also may occur.

Some studies in adults and children have linked organophosphate exposure to lymphoma and leukemia. Home pesticide use overall has been linked to childhood cancers such as soft tissue sarcomas, leukemias, and cancer of the brain. The results of these studies are controversial. No studies have definitively linked these exposures with cancer because the exposure is not measured and people usually are not exposed to just the one pesticide being studied.

There is no evidence to suggest that organophosphate exposure affects human reproduction or development.

Most organophosphates do not cause reproductive or developmental problems in animals, but some cause lower birth weights or more death in newborn animals.

Organophosphate insecticides are highly toxic to insects but are relatively less so to humans and domestic animals. Although specific organophosphates, such as triorthocresyl phosphate, nipafox, and trichlorofon compounds, are obviously neurotoxic, the situation is less clear for other organophosphate pesticides. At risk are factory workers involved in the production of these compounds and agricultural workers who use them to spray crops. The occurrence of organophosphate poisoning in the United States is low, although epidemics have been reported in some Third World countries. It should be noted, however, that 3011 pesticide-related cases were reported in California and Washington between 1993 and 1998. California accounts for approximately one third of the agricultural workforce in the United States. Organophosphates are absorbed through the skin and respiratory tracts. In addition, small amounts may be ingested by eating foods that have been sprayed with organophosphates.

References

- Stokstad E (June 2007). "Species conservation. Can the bald eagle still soar after it is delisted?". Science. 316 (5832): 1689–90. doi:10.1126/science.316.5832.1689. PMID 17588911
- Michaels D (2008). Doubt is Their Product: How Industry's Assault on Science Threatens Your Health. New York: Oxford University Press. ISBN 978-0-19-530067-3
- Murray B. Isman "Botanical Insecticides, Deterrents, And Repellents In Modern Agriculture And An Increasingly Regulated World" Annual Review Of Entomology Volume 51, pp. 45-66. doi:10.1146/annurev.ento.51.110104.151146
- Elizabeth Norton Lasley. "Having Their Toxins and Eating Them Too Study of the natural sources of many animals' chemical defenses is providing new insights into nature's medicine chest". Oxford Journals. pp. 945–950. Retrieved 2015-05-06
- Davis, Frederick Rowe (2014). Banned : a history of pesticides and the science of toxicology. [S.l.]: Yale University Press. p. 26. ISBN 978-0300205176. Retrieved 25 July 2017
- DDT and Its Derivatives: Environmental Aspects, Environmental Health Criteria monograph No. 83, Geneva: World Health Organization, ISBN 92-4-154283-7
- Oldroyd, B.P. (2007). "What›s Killing American Honey Bees?". PLoS Biology. 5 (6): e168. doi:10.1371/journal.pbio.0050168. PMC 1892840. PMID 17564497
- Metcalf, Robert L. (2002). "Insect Control". Ullmann's Encyclopedia of Industrial Chemistry. Ullmann's Encyclopedia of Industrial Chemistry. Wiley-VCH. doi:10.1002/14356007.a14_263. ISBN 3527306730

Chapter 4

Pesticides

The substances that stop the growth of pests are termed as pesticides. Some of the various types of pesticides are herbicide, molluscicide, acaricide, piscicide and dicofol. Herbicides are the substances which are used to control the growth of unwanted plants. The chapter closely examines these types of pesticides to provide an extensive understanding of the subject.

Pesticide

Pesticide is any toxic substance used to kill animals, fungi, or plants that cause economic damage to crop or ornamental plants or are hazardous to the health of domestic animals or humans. All pesticides interfere with normal metabolic processes in the pest organism and often are classified according to the type of organism they are intended to control.

The use of certain pesticides is controversial. The insecticide DDT, for example, which came into wide use in the 1940s, was later heavily restricted in the United States and elsewhere because of its adverse affects on the health of the environment, wildlife, and humans. In the early 21st century the use of neonicotinoids was highly restricted in some countries, including throughout the entire European Union, because of the possible involvement of those insecticides in the decline of honeybee populations.

Active Ingredients

The active ingredient, usually the only component of the formulation listed on the pesticide label, is by nature biologically and chemically active against a target pest, be it an insect, weed or fungus. By definition these chemicals kill living things.

Contaminants and Impurities

Contaminants and impurities are often a part of the pesticide product and responsible for product hazards. Dioxin and DDT have been identified as contaminants, which have not been purposefully added but are a function of the production process.

Metabolites

Metabolites are breakdown products that form when a pesticide is used in the environment and mixes with air, water, soil or living organisms. Often the metabolite is more hazardous than the parent pesticide.

Inert Ingredients

If you were to go to your local hardware store and take a look at the label on a can of ant and roach killer, the contents might read something like this, "5% Permethrin, 95% Inert Ingredients." After reading the label, you may wonder what makes up the other 95%. The fact is, the manufacturer doesn't have to tell you. Currently, under the Federal Insecticide, Fungicide and Rodenticide Act (FIFRA), pesticide manufacturers are only required to list the active ingredients in a pesticide, leaving consumers and applicators unaware of the possible toxics present in the inert ingredients of pesticide products they are using, unless the EPA administrator determines that the chemical poses a public health threat. Pesticide manufacturers argue they cannot release information on inert ingredients because they are trade secrets, and if released, their products could be duplicated. Quite often inert ingredients constitute over 95% of the pesticide product. Inert ingredients are mixed into pesticides products as a carrier or sticking agent, and are often as toxic as the active ingredient.

Alternatives

Alternatives to pesticides are available and include methods of cultivation, use of biological pest controls (such as pheromones and microbial pesticides), genetic engineering, and methods of interfering with insect breeding. Application of composted yard waste has also been used as a way of controlling pests. These methods are becoming increasingly popular and often are safer than traditional chemical pesticides. In addition, EPA is registering reduced-risk conventional pesticides in increasing numbers.

Cultivation practices include polyculture (growing multiple types of plants), crop rotation, planting crops in areas where the pests that damage them do not live, timing planting according to when pests will be least problematic, and use of trap crops that attract pests away from the real crop. Trap crops have successfully controlled pests in some commercial agricultural systems while reducing pesticide usage;however, in many other systems, trap crops can fail to reduce pest densities at a commercial scale, even when the trap crop works in controlled experiments. In the U.S., farmers have had success controlling insects by spraying with hot water at a cost that is about the same as pesticide spraying.

Release of other organisms that fight the pest is another example of an alternative to pesticide use. These organisms can include natural predators or parasites of the pests.

Biological pesticides based on entomopathogenic fungi, bacteria and viruses cause disease in the pest species can also be used.

Interfering with insects' reproduction can be accomplished by sterilizing males of the target species and releasing them, so that they mate with females but do not produce offspring. This technique was first used on the screwworm fly in 1958 and has since been used with the medfly, the tsetse fly, and the gypsy moth. However, this can be a costly, time consuming approach that only works on some types of insects.

Push Pull Strategy

The term "push-pull" was established in 1987 as an approach for integrated pest management (IPM). This strategy uses a mixture of behavior-modifying stimuli to manipulate the distribution and abundance of insects. "Push" means the insects are repelled or deterred away from whatever resource that is being protected. "Pull" means that certain stimuli (semiochemical stimuli, pheromones, food additives, visual stimuli, genetically altered plants, etc.) are used to attract pests to trap crops where they will be killed. There are numerous different components involved in order to implement a Push-Pull Strategy in IPM.

Many case studies testing the effectiveness of the push-pull approach have been done across the world. The most successful push-pull strategy was developed in Africa for subsistence farming. Another successful case study was performed on the control of *Helicoverpa* in cotton crops in Australia. In Europe, the Middle East, and the United States, push-pull strategies were successfully used in the controlling of *Sitona lineatus* in bean fields.

Some advantages of using the push-pull method are less use of chemical or biological materials and better protection against insect habituation to this control method. Some disadvantages of the push-pull strategy is that if there is a lack of appropriate knowledge of behavioral and chemical ecology of the host-pest interactions then this method becomes unreliable. Furthermore, because the push-pull method is not a very popular method of IPM operational and registration costs are higher.

Effectiveness

Some evidence shows that alternatives to pesticides can be equally effective as the use of chemicals. For example, Sweden has halved its use of pesticides with hardly any reduction in crops. In Indonesia, farmers have reduced pesticide use on rice fields by 65% and experienced a 15% crop increase. A study of Maize fields in northern Florida found that the application of composted yard waste with high carbon to nitrogen ratio to agricultural fields was highly effective at reducing the population of plant-parasitic nematodes and increasing crop yield, with yield increases ranging from 10% to 212%; the observed effects were long-term, often not appearing until the third season of the study.

However, pesticide resistance is increasing. In the 1940s, U.S. farmers lost only 7% of their crops to pests. Since the 1980s, loss has increased to 13%, even though more pesticides are being used. Between 500 and 1,000 insect and weed species have developed pesticide resistance since 1945.

Types

Pesticides are often referred to according to the type of pest they control. Pesticides can also be considered as either biodegradable pesticides, which will be broken down by microbes and other living beings into harmless compounds, or persistent pesticides, which may take months or years before they are broken down: it was the persistence of DDT, for example, which led to its accumulation in the food chain and its killing of birds of prey at the top of the food chain. Another way to think about pesticides is to consider those that are chemical pesticides are derived from a common source or production method.

Insecticides

Neonicotinoids are a class of neuro-active insecticides chemically similar to nicotine. Imidacloprid, of the neonicotanoid family, is the most widely used insecticide in the world. In the late 1990s neonicotinoids came under increasing scrutiny over their environmental impact and were linked in a range of studies to adverse ecological effects, including honey-bee colony collapse disorder (CCD) and loss of birds due to a reduction in insect populations. In 2013, the European Union and a few non EU countries restricted the use of certain neonicotinoids.

Organophosphate and carbamate insecticides have a similar mode of action. They affect the nervous system of target pests (and non-target organisms) by disrupting acetylcholinesterase activity, the enzyme that regulates acetylcholine, at nerve synapses. This inhibition causes an increase in synaptic acetylcholine and over-stimulation of the parasympathetic nervous system. Many of these insecticides, first developed in the mid 20th century, are very poisonous. Although commonly used in the past, many older chemicals have been removed from the market due to their health and environmental effects (*e.g.* DDT, chlordane, and toxaphene). However, many organophosphates are not persistent in the environment.

Pyrethroid insecticides were developed as a synthetic version of the naturally occurring pesticide pyrethrin, which is found in chrysanthemums. They have been modified to increase their stability in the environment. Some synthetic pyrethroids are toxic to the nervous system.

Herbicides

A number of sulfonylureas have been commercialized for weed control, including: amidosulfuron, flazasulfuron, metsulfuron-methyl, rimsulfuron, sulfometuron-methyl, terbacil, nicosulfuron, and triflusulfuron-methyl. These are broad-spectrum herbicides that kill plants weeds or pests by inhibiting the enzyme acetolactate synthase. In the 1960s, more than 1 kg/ha (0.89 lb/acre) crop protection chemical was typically applied, while sulfonylureates allow as little as 1% as much material to achieve the same effect.

Biopesticides

Biopesticides are certain types of pesticides derived from such natural materials as animals, plants, bacteria, and certain minerals. For example, canola oil and baking soda have pesticidal applications and are considered biopesticides. Biopesticides fall into three major classes:

- Microbial pesticides which consist of bacteria, entomopathogenic fungi or viruses (and sometimes includes the metabolites that bacteria or fungi produce). Entomopathogenic nematodes are also often classed as microbial pesticides, even though they are multi-cellular.
- Biochemical pesticides or herbal pesticides are naturally occurring substances that control (or monitor in the case of pheromones) pests and microbial diseases.
- Plant-incorporated protectants (PIPs) have genetic material from other species incorporated into their genetic material (i.e. GM crops). Their use is controversial, especially in many European countries.

Classified by Type of Pest

Pesticides that are related to the type of pests are:

Type	Action
Algicides	Control algae in lakes, canals, swimming pools, water tanks, and other sites
Antifouling agents	Kill or repel organisms that attach to underwater surfaces, such as boat bottoms
Antimicrobials	Kill microorganisms (such as bacteria and viruses)
Attractants	Attract pests (for example, to lure an insect or rodent to a trap). (However, food is not considered a pesticide when used as an attractant.)
Biopesticides	Biopesticides are certain types of pesticides derived from such natural materials as animals, plants, bacteria, and certain minerals
Biocides	Kill microorganisms
Disinfectants and sanitizers	Kill or inactivate disease-producing microorganisms on inanimate objects
Fungicides	Kill fungi (including blights, mildews, molds, and rusts)
Fumigants	Produce gas or vapor intended to destroy pests in buildings or soil
Herbicides	Kill weeds and other plants that grow where they are not wanted
Insecticides	Kill insects and other arthropods
Miticides	Kill mites that feed on plants and animals

Microbial pesticides	Microorganisms that kill, inhibit, or out compete pests, including insects or other microorganisms
Molluscicides	Kill snails and slugs
Nematicides	Kill nematodes (microscopic, worm-like organisms that feed on plant roots)
Ovicides	Kill eggs of insects and mites
Pheromones	Biochemicals used to disrupt the mating behavior of insects
Repellents	Repel pests, including insects (such as mosquitoes) and birds
Rodenticides	Control mice and other rodents

Further Types

The term pesticide also include these substances:

Defoliants: Cause leaves or other foliage to drop from a plant, usually to facilitate harvest.

Desiccants: Promote drying of living tissues, such as unwanted plant tops.

Insect growth regulators: Disrupt the molting, maturity from pupal stage to adult, or other life processes of insects.

Plant growth regulators: Substances (excluding fertilizers or other plant nutrients) that alter the expected growth, flowering, or reproduction rate of plants.

Wood preservatives: They are used to make wood resistant to insects, fungus, and other pests.

Residue

Pesticide residue refers to the pesticides that may remain on or in food after they are applied to food crops. The maximum allowable levels of these residues in foods are often stipulated by regulatory bodies in many countries. Regulations such as pre-harvest intervals also often prevent harvest of crop or livestock products if recently treated in order to allow residue concentrations to decrease over time to safe levels before harvest. Exposure of the general population to these residues most commonly occurs through consumption of treated food sources, or being in close contact to areas treated with pesticides such as farms or lawns.

Many of these chemical residues, especially derivatives of chlorinated pesticides, exhibit bioaccumulation which could build up to harmful levels in the body as well as in the environment. Persistent chemicals can be magnified through the food chain and have been detected in products ranging from meat, poultry, and fish, to vegetable oils, nuts, and various fruits and vegetables.

Because of the toxicity of pesticides in concentrated form and because of the frequency of exposure, the most serious human health effects are found among agricultural and production workers. Long-term and chronic health effects occur as the chemicals are ingested and inhaled. Bioaccumulation (the buildup of toxic materials in tissues) is evident in fish and birds as well as in humans.

Two different sets of problems are posed by the occurrence of persistent pesticides in the environment: (1) localized problems, tending to be acute, leading to recognizable effects with assignable origins, which can be dealt with if successfully sorted out and if the will and powers exist; and (2) more widespread problems, tending to be inferential with postulated effects, due to the universal presence of the materials in question. There is, for example, growing evidence from the amount of pesticide residues found in specimens of affected species as well as in the animals, fish, invertebrates, or plankton they feed upon, that there is a process of pesticide concentration in the food chains. As small amounts of persistent pesticides become more and more widely spread throughout the entire natural environment, they are absorbed by low forms of life. Where large numbers of these species serve as food for higher animals, some of the total pesticide remains in the eater. This can lead to a certain concentration of pesticides in a form of life at the end of the food chain. Eventually lethal doses may be reached for certain populations, or the species may be reduced because of adverse effects on reproduction or behaviour.

The importance of changes in the environment is that they may be irreversible: species may disappear altogether. This will cause ecological imbalance - certain forms of life whose numbers are controlled by the disappearing species will no longer be held in check. Thus an apparently harmless species may become a pest. Another result of reduction of species diversity may be the loss of the genetic possibilities that each organism disturbance of the selective pressures of the environment will reduce the chances of future development of desired plants and animals. Furthermore, as humans are at the top of food chains, they too tend to concentrate residues in their bodies with as yet unknown effects on them. While perhaps sufficient attention has been directed to acute toxicity problems, too little attention has been paid to the effect of long term ingestion by human beings of small amounts of these chemicals.

A serious dilemma is that the cheapest pesticides are the organochlorines, known to accumulate in human fat. The substitution of organophosphorus pesticides brings both higher cost and a need for more carefully trained operators, both of which pose problems.

Herbicide

A herbicide is a chemical substance used to control or manipulate undesirable vegetation, especially weeds. Herbicides are extensively used in gardening, farming, and landscape turf management.

Herbicides are classified into two categories: selective and non-selective. Selective herbicides kill specific unwanted plants while leaving desirable vegetation relatively unharmed. Non-selective herbicides (total weed killers) kill all or most plant species.

A herbicide can be applied directly to the plant, applied to the soil, or sprayed onto the foliage. Herbicides are applied before, during, or after crop planting in row-crop farming to maximize crop production by diminishing the development of unwanted plants. Herbicides are also applied in ponds and lakes to control aquatic plants, in forests to prepare logged areas for replanting, and to golf courses, lawns, parks, and other areas to clear out unwanted vegetation.

Besides selective and non-selective classifications, a herbicide can also be categorized according to three other characteristics:

- Persistence - How long the herbicide remains potent.
- Mechanism of action - How it works.
- Means of uptake - How the plants will absorb it (e.g., through the roots, aboveground foliage, etc.)

A herbicide's effectiveness is strongly influenced by its toxic mode of action and the application method. Herbicides can act by inhibiting a plant's amino acid production, growth, photosynthesis, cell division, or by mimicking natural auxin hormones to cause deformities. Most modern herbicides are synthetic mimics of a natural plant's hormones that obstruct the target plant's growth.

Some plants such as the tree of heaven and juglans (walnuts) produce their own natural herbicide. Organic herbicides are useful and are commonly used in organic gardens, but they are less effective and more costly than synthetic herbicides because they based on natural materials.

For difficult cases, a combination of several herbicides is recommended when dealing with herbicide resistance.

Terminology

Herbicides are classified/grouped in various ways e.g. according to the activity, timing of application, method of application, mechanism of action, chemical family. This gives rise to a considerable level of terminology related to herbicides and their use.

Intended Outcome

- Control is the destruction of unwanted weeds, or the damage of them to the point where they are no longer competitive with the crop.

- Suppression is incomplete control still providing some economic benefit, such as reduced competition with the crop.
- Crop safety, for selective herbicides, is the relative absence of damage or stress to the crop. Most selective herbicides cause some visible stress to crop plants.
- Defoliant, similar to herbicides, but designed to remove foliage (leaves) rather than kill the plant.

Selectivity

- Selective herbicides control or suppress certain plants without affecting the growth of other plants species. Selectivity may be due to translocation, differential absorption, physical (morphological) or physiological differences between plant species. 2,4-D, mecoprop, dicamba control many broadleaf weeds but remain ineffective against turfgrasses.
- Non-selective herbicides are not specific in acting against certain plant species and control all plant material with which they come into contact. They are used to clear industrial sites, waste ground, railways and railway embankments. Paraquat, glufosinate, glyphosate are non-selective herbicides.

Timing of Application

- Preplant: Preplant herbicides are nonselective herbicides applied to soil before planting. Some preplant herbicides may be mechanically incorporated into the soil. The objective for incorporation is to prevent dissipation through photodecomposition and/or volatility. The herbicides kill weeds as they grow through the herbicide treated zone. Volatile herbicides have to be incorporated into the soil before planting the pasture. Agricultural crops grown in soil treated with a preplant herbicide include tomatoes, corn, soybeans and strawberries. Soil fumigants like metam-sodium and dazomet are in use as preplant herbicides.
- Preemergence: Preemergence herbicides are applied before the weed seedlings emerge through the soil surface. Herbicides do not prevent weeds from germinating but they kill weeds as they grow through the herbicide treated zone by affecting the cell division in the emerging seedling. Dithopyr and pendimethalin are preemergence herbicides. Weeds that have already emerged before application or activation are not affected by pre-herbicides as their primary growing point escapes the treatment.
- Postemergence: These herbicides are applied after weed seedlings have emerged through the soil surface. They can be foliar or root absorbed, selective or non-selective, contact or systemic. Application of these herbicides is avoided during rain because the problem of being washed off to the soil makes it ineffective. 2,4-D is a selective, systemic, foliar absorbed postemergence herbicide.

Method of Application

- Soil applied: Herbicides applied to the soil are usually taken up by the root or shoot of the emerging seedlings and are used as preplant or preemergence treatment. Several factors influence the effectiveness of soil-applied herbicides. Weeds absorb herbicides by both passive and active mechanism. Herbicide adsorption to soil colloids or organic matter often reduces its amount available for weed absorption. Positioning of herbicide in correct layer of soil is very important, which can be achieved mechanically and by rainfall. Herbicides on the soil surface are subjected to several processes that reduce their availability. Volatility and photolysis are two common processes that reduce the availability of herbicides. Many soil applied herbicides are absorbed through plant shoots while they are still underground leading to their death or injury. EPTC and trifluralin are soil applied herbicides.
- Foliar applied: These are applied to portion of the plant above the ground and are absorbed by exposed tissues. These are generally postemergence herbicides and can either be translocated (systemic) throughout the plant or remain at specific site (contact). External barriers of plants like cuticle, waxes, cell wall etc. affect herbicide absorption and action. Glyphosate, 2,4-D and dicamba are foliar applied herbicide.

Persistence

- Residual activity: An herbicide is described as having low residual activity if it is neutralized within a short time of application (within a few weeks or months) - typically this is due to rainfall, or by reactions in the soil. An herbicide described as having high residual activity will remain potent for a long term in the soil. For some compounds, the residual activity can leave the ground almost permanently barren.

Mechanism of Action

Herbicides are often classified according to their site of action, because as a general rule, herbicides within the same site of action class will produce similar symptoms on susceptible plants. Classification based on site of action of herbicide is comparatively better as herbicide resistance management can be handled more properly and effectively. Classification by mechanism of action (MOA) indicates the first enzyme, protein, or biochemical step affected in the plant following application.

List of Mechanisms found in Modern Herbicides

- ACCase inhibitors: Acetyl coenzyme A carboxylase (ACCase) is part of the first step of lipid synthesis. Thus, ACCase inhibitors affect cell membrane production in the meristems of the grass plant. The ACCases of grasses are sensitive to these herbicides, whereas the ACCases of dicot plants are not.

- ALS inhibitors: the acetolactate synthase (ALS) enzyme (also known as acetohydroxyacid synthase, or AHAS) is the first step in the synthesis of the branched-chain amino acids (valine, leucine, and isoleucine). These herbicides slowly starve affected plants of these amino acids, which eventually leads to inhibition of DNA synthesis. They affect grasses and dicots alike. The ALS inhibitor family includes various sulfonylureas (SUs) (such as Flazasulfuron and Metsulfuron-methyl), imidazolinones (IMIs), triazolopyrimidines (TPs), pyrimidinyl oxybenzoates (POBs), and sulfonylamino carbonyl triazolinones(SCTs). The ALS biological pathway exists only in plants and not animals, thus making the ALS-inhibitors among the safest herbicides.
- EPSPS inhibitors: Enolpyruvylshikimate 3-phosphate synthase enzyme (EPSPS) is used in the synthesis of the amino acids tryptophan, phenylalanine and tyrosine. They affect grasses and dicots alike. Glyphosate (Roundup) is a systemic EPSPS inhibitor inactivated by soil contact.
- Synthetic auxins inaugurated the era of organic herbicides. They were discovered in the 1940s after a long study of the plant growth regulator auxin. Synthetic auxins mimic this plant hormone. They have several points of action on the cell membrane, and are effective in the control of dicot plants. 2,4-D is a synthetic auxin herbicide.
- Photosystem II inhibitors reduce electron flow from water to $NADP^+$ at the photochemical step in photosynthesis. They bind to the Qb site on the D1 protein, and prevent quinone from binding to this site. Therefore, this group of compounds causes electrons to accumulate on chlorophyll molecules. As a consequence, oxidation reactions in excess of those normally tolerated by the cell occur, and the plant dies. The triazine herbicides (including atrazine) and urea derivatives (diuron) are photosystem II inhibitors.
- Photosystem I inhibitors steal electrons from the normal pathway through FeS to Fdx to $NADP^+$ leading to direct discharge of electrons on oxygen. As a result, reactive oxygen species are produced and oxidation reactions in excess of those normally tolerated by the cell occur, leading to plant death. Bipyridinium herbicides (such as diquat and paraquat) inhibit the FeS to Fdx step of that chain, while diphenyl ether herbicides (such as nitrofen, nitrofluorfen, and acifluorfen) inhibit the Fdx to $NADP^+$ step.
- HPPD inhibitors inhibit 4-Hydroxyphenylpyruvate dioxygenase, which are involved in tyrosine breakdown. Tyrosine breakdown products are used by plants to make carotenoids, which protect chlorophyll in plants from being destroyed by sunlight. If this happens, the plants turn white due to complete loss of chlorophyll, and the plants die. Mesotrione and sulcotrione are herbicides in this class; a drug, nitisinone, was discovered in the course of developing this class of herbicides.

Herbicide Group

One of the most important methods for preventing, delaying, or managing resistance is to reduce the reliance on a single herbicide mode of action. To do this, farmers must know the mode of action for the herbicides they intend to use, but the relatively complex nature of plant biochemistry makes this difficult to determine. Attempts were made to simplify the understanding of herbicide mode of action by developing a classification system that grouped herbicides by mode of action. Eventually the Herbicide Resistance Action Committee (HRAC) and the Weed Science Society of America (WSSA) developed a classification system. The WSSA and HRAC systems differ in the group designation. Groups in the WSSA and the HRAC systems are designated by numbers and letters, respectively. The goal for adding the "Group" classification and mode of action to the herbicide product label is to provide a simple and practical approach to deliver the information to users. This information will make it easier to develop educational material that is consistent and effective. It should increase user's awareness of herbicide mode of action and provide more accurate recommendations for resistance management.Another goal is to make it easier for users to keep records on which herbicide mode of actions are being used on a particular field from year to year.

Chemical Family

Detailed investigations on chemical structure of the active ingredients of the registered herbicides showed that some moieties (moiety is a part of a molecule that may include either whole functional groups or parts of functional groups as substructures; a functional group has similar chemical properties whenever it occurs in different compounds) have the same mechanisms of action. According to Forouzesh *et al.* 2015, these moieties have been assigned to the names of chemical families and active ingredients are then classified within the chemical families accordingly. Knowing about herbicide chemical family grouping could serve as a short-term strategy for managing resistance to site of action.

Use and Application

Herbicides being sprayed from the spray arms of a tractor in North Dakota.

Most herbicides are applied as water-based sprays using ground equipment. Ground equipment varies in design, but large areas can be sprayed using self-propelled sprayers equipped with long booms, of 60 to 120 feet (18 to 37 m) with spray nozzles spaced every 20–30 inches (510–760 mm) apart. Towed, handheld, and even horse-drawn sprayers are also used. On large areas, herbicides may also at times be applied aerially using helicopters or airplanes, or through irrigation systems (known as chemigation).

A further method of herbicide application developed around 2010, involves ridding the soil of its active weed seed bank rather than just killing the weed. This can successfully treat annual plants but not perennials. Researchers at the Agricultural Research Service found that the application of herbicides to fields late in the weeds› growing season greatly reduces their seed production, and therefore fewer weeds will return the following season. Because most weeds are annuals, their seeds will only survive in soil for a year or two, so this method will be able to destroy such weeds after a few years of herbicide application.

Weed-wiping may also be used, where a wick wetted with herbicide is suspended from a boom and dragged or rolled across the tops of the taller weed plants. This allows treatment of taller grassland weeds by direct contact without affecting related but desirable shorter plants in the grassland sward beneath. The method has the benefit of avoiding spray drift. In Wales, a scheme offering free weed-wiper hire was launched in 2015 in an effort to reduce the levels of MCPA in water courses.

Misuse and Misapplication

Herbicide volatilisation or spray drift may result in herbicide affecting neighboring fields or plants, particularly in windy conditions. Sometimes, the wrong field or plants may be sprayed due to error.

Health and Environmental Effects

Herbicides have widely variable toxicity in addition to acute toxicity arising from ingestion of a significant quantity rapidly, and chronic toxicity arising from environmental and occupational exposure over long periods. Much public suspicion of herbicides revolves around a confusion between valid statements of *acute* toxicity as opposed to equally valid statements of lack of *chronic* toxicity at the recommended levels of usage. For instance, while glyphosate formulations with talloamine *adjuvants* are acutely toxic, their use was found to be uncorrelated with any health issues like cancer in a massive US Department of Health study on 90,000 members of farmer families for over a period of 23 years . That is, the study shows lack of chronic toxicity, but cannot question the herbicide's acute toxicity.

Some herbicides cause a range of health effects ranging from skin rashes to death. The pathway of attack can arise from intentional or unintentional direct consumption, improper application resulting in the herbicide coming into direct contact with people or wildlife, inhalation of aerial sprays, or food consumption prior to the labelled preharvest interval. Under some conditions, certain herbicides can be transported via leaching or surface runoff to contaminate groundwater or distant surface water sources. Generally, the conditions that promote herbicide transport include intense storm events (particularly shortly after application) and soils with limited capacity to adsorb or retain the herbicides. Herbicide properties that increase likelihood of transport include persistence (resistance to degradation) and high water solubility.

Phenoxy herbicides are often contaminated with dioxins such as TCDD; research has suggested such contamination results in a small rise in cancer risk after occupational exposure to these herbicides. Triazine exposure has been implicated in a likely relationship to increased risk of breast cancer, although a causal relationship remains unclear.

Herbicide manufacturers have at times made false or misleading claims about the safety of their products. Chemical manufacturer Monsanto Company agreed to change its advertising after pressure from New York attorney general Dennis Vacco; Vacco complained about misleading claims that its spray-on glyphosate-based herbicides, including Roundup, were safer than table salt and "practically non-toxic" to mammals, birds, and fish (though proof that this was ever said is hard to find). Roundup is toxic and has resulted in death after being ingested in quantities ranging from 85 to 200 ml, although it has also been ingested in quantities as large as 500 ml with only mild or moderate symptoms. The manufacturer of Tordon 101 (Dow AgroSciences, owned by the Dow Chemical Company) has claimed Tordon 101 has no effects on animals and insects, in spite of evidence of strong carcinogenic activity of the active ingredient Picloram in studies on rats.

The risk of Parkinson's disease has been shown to increase with occupational exposure to herbicides and pesticides. The herbicide paraquat is suspected to be one such factor.

All commercially sold, organic and nonorganic herbicides must be extensively tested prior to approval for sale and labeling by the Environmental Protection Agency. However, because of the large number of herbicides in use, concern regarding health effects is significant. In addition to health effects caused by herbicides themselves, commercial herbicide mixtures often contain other chemicals, including inactive ingredients, which have negative impacts on human health.

Ecological Effects

Commercial herbicide use generally has negative impacts on bird populations, although the impacts are highly variable and often require field studies to predict accurately. Laboratory studies have at times overestimated negative impacts on birds due to toxicity, predicting serious problems that were not observed in the field. Most observed effects are due not to toxicity, but to habitat changes and the decreases in abundance of species on which birds rely for food or shelter. Herbicide use in silviculture, used to favor certain types of growth following clearcutting, can cause significant drops in bird populations. Even when herbicides which have low toxicity to birds are used, they decrease the abundance of many types of vegetation on which the birds rely. Herbicide use in agriculture in Britain has been linked to a decline in seed-eating bird species which rely on the weeds killed by the herbicides. Heavy use of herbicides in neotropical agricultural areas has been one of many factors implicated in limiting the usefulness of such agricultural land for wintering migratory birds.

Frog populations may be affected negatively by the use of herbicides as well. While some studies have shown that atrazine may be a teratogen, causing demasculinization in male frogs, the U.S. Environmental Protection Agency (EPA) and its independent Scientific Advisory Panel (SAP) examined all available studies on this topic and concluded that "atrazine does not adversely affect amphibian gonadal development based on a review of laboratory and field studies."

Scientific Uncertainty of Full Extent of Herbicide Effects

The health and environmental effects of many herbicides is unknown, and even the scientific community often disagrees on the risk. For example, a 1995 panel of 13 scientists reviewing studies on the carcinogenicity of 2,4-D had divided opinions on the likelihood 2,4-D causes cancer in humans. As of 1992, studies on phenoxy herbicides were too few to accurately assess the risk of many types of cancer from these herbicides, even though evidence was stronger that exposure to these herbicides is associated with increased risk of soft tissue sarcoma and non-Hodgkin lymphoma. Furthermore, there is some suggestion that herbicides can play a role in sex reversal of certain organisms that experience temperature-dependent sex determination, which could theoretically alter sex ratios.

Resistance

Weed resistance to herbicides has become a major concern in crop production worldwide. Resistance to herbicides is often attributed to lack of rotational programmes of herbicides and to continuous applications of herbicides with the same sites of action. Thus, a true understanding of the sites of action of herbicides is essential for strategic planning of herbicide-based weed control.

Plants have developed resistance to atrazine and to ALS-inhibitors, and more recently, to glyphosate herbicides. Marestail is one weed that has developed glyphosate resistance. Glyphosate-resistant weeds are present in the vast majority of soybean, cotton and corn farms in some U.S. states. Weeds that can resist multiple other herbicides are spreading. Few new herbicides are near commercialization, and none with a molecular mode of action for which there is no resistance. Because most herbicides could not kill all weeds, farmers rotated crops and herbicides to stop resistant weeds. During its initial years, glyphosate was not subject to resistance and allowed farmers to reduce the use of rotation.

A family of weeds that includes waterhemp (Amaranthus rudis) is the largest concern. A 2008-9 survey of 144 populations of waterhemp in 41 Missouri counties revealed glyphosate resistance in 69%. Weeds from some 500 sites throughout Iowa in 2011 and 2012 revealed glyphosate resistance in approximately 64% of waterhemp samples. The use of other killers to target "residual" weeds has become common, and may be sufficient to have stopped the spread of resistance From 2005 through 2010 researchers discovered 13 different weed species that had developed resistance to glyphosate. But since then only two more have been discovered. Weeds resistant to multiple herbicides with completely different biological action modes are on the rise. In Missouri, 43% of samples were resistant to two different herbicides; 6% resisted three; and 0.5% resisted four. In Iowa 89% of waterhemp samples resist two or more herbicides, 25% resist three, and 10% resist five.

For southern cotton, herbicide costs has climbed from between $50 and $75 per hectare a few years ago to about $370 per hectare in 2013. Resistance is contributing to a massive shift away from growing cotton; over the past few years, the area planted with cotton has declined by 70% in Arkansas and by 60% in Tennessee. For soybeans in Illinois, costs have risen from about $25 to $160 per hectare.

Biochemistry of Resistance

Resistance to herbicides can be based on one of the following biochemical mechanisms:

- Target-site resistance: This is due to a reduced (or even lost) ability of the herbicide to bind to its target protein. The effect usually relates to an enzyme with a crucial function in a metabolic pathway, or to a component of an electron-transport system. Target-site resistance may also be caused by an overexpression of the target enzyme (via gene amplification or changes in a gene promoter).
- Non-target-site resistance: This is caused by mechanisms that reduce the amount of herbicidal active compound reaching the target site. One important mechanism is an enhanced metabolic detoxification of the herbicide in the weed, which leads to insufficient amounts of the active substance reaching the target site. A reduced uptake and translocation, or sequestration of the herbicide, may also result in an insufficient herbicide transport to the target site.
- Cross-resistance: In this case, a single resistance mechanism causes resistance to several herbicides. The term target-site cross-resistance is used when the herbicides bind to the same target site, whereas non-target-site cross-resistance is due to a single non-target-site mechanism (e.g., enhanced metabolic detoxification) that entails resistance across herbicides with different sites of action.

- Multiple resistance: In this situation, two or more resistance mechanisms are present within individual plants, or within a plant population.

Resistance Management

Worldwide experience has been that farmers tend to do little to prevent herbicide resistance developing, and only take action when it is a problem on their own farm or neighbor's. Careful observation is important so that any reduction in herbicide efficacy can be detected. This may indicate evolving resistance. It is vital that resistance is detected at an early stage as if it becomes an acute, whole-farm problem, options are more limited and greater expense is almost inevitable. Table below lists factors which enable the risk of resistance to be assessed. An essential pre-requisite for confirmation of resistance is a good diagnostic test. Ideally this should be rapid, accurate, cheap and accessible. Many diagnostic tests have been developed, including glasshouse pot assays, petri dish assays and chlorophyll fluorescence. A key component of such tests is that the response of the suspect population to a herbicide can be compared with that of known susceptible and resistant standards under controlled conditions. Most cases of herbicide resistance are a consequence of the repeated use of herbicides, often in association with crop monoculture and reduced cultivation practices. It is necessary, therefore, to modify these practices in order to prevent or delay the onset of resistance or to control existing resistant populations. A key objective should be the reduction in selection pressure. An integrated weed management (IWM) approach is required, in which as many tactics as possible are used to combat weeds. In this way, less reliance is placed on herbicides and so selection pressure should be reduced.

Optimising herbicide input to the economic threshold level should avoid the unnecessary use of herbicides and reduce selection pressure. Herbicides should be used to their greatest potential by ensuring that the timing, dose, application method, soil and climatic conditions are optimal for good activity. In the UK, partially resistant grass weeds such as *Alopecurus myosuroides* (blackgrass) and *Avena* spp. (wild oat) can often be controlled adequately when herbicides are applied at the 2-3 leaf stage, whereas later applications at the 2-3 tiller stage can fail badly. Patch spraying, or applying herbicide to only the badly infested areas of fields, is another means of reducing total herbicide use.

Table: Agronomic factors influencing the risk of herbicide resistance development.

Factor	Low risk	High risk
Cropping system	Good rotation	Crop monoculture
Cultivation system	Annual ploughing	Continuous minimum tillage
Weed control	Cultural only	Herbicide only
Herbicide use	Many modes of action	Single modes of action
Control in previous years	Excellent	Poor
Weed infestation	Low	High
Resistance in vicinity	Unknown	Common

Approaches to Treating Resistant Weeds

Alternative Herbicides

When resistance is first suspected or confirmed, the efficacy of alternatives is likely to be the first consideration. The use of alternative herbicides which remain effective on resistant populations can be a successful strategy, at least in the short term. The effectiveness of alternative herbicides will be highly dependent on the extent of cross-resistance. If there is resistance to a single group of herbicides, then the use of herbicides from other groups may provide a simple and effective solution, at least in the short term. For example, many triazine-resistant weeds have been readily controlled by the use of alternative herbicides such as dicamba or glyphosate. If resistance extends to more than one herbicide group, then choices are more limited. It should not be assumed that resistance will automatically extend to all herbicides with the same mode of action, although it is wise to assume this until proved otherwise. In many weeds the degree of cross-resistance between the five groups of ALS inhibitors varies considerably. Much will depend on the resistance mechanisms present, and it should not be assumed that these will necessarily be the same in different populations of the same species. These differences are due, at least in part, to the existence of different mutations conferring target site resistance. Consequently, selection for different mutations may result in different patterns of cross-resistance. Enhanced metabolism can affect even closely related herbicides to differing degrees. For example, populations of *Alopecurus myosuroides* (blackgrass) with an enhanced metabolism mechanism show resistance to pendimethalin but not to trifluralin, despite both being dinitroanilines. This is due to differences in the vulnerability of these two herbicides to oxidative metabolism. Consequently, care is needed when trying to predict the efficacy of alternative herbicides.

Mixtures and Sequences

The use of two or more herbicides which have differing modes of action can reduce the selection for resistant genotypes. Ideally, each component in a mixture should:

- Be active at different target sites.
- Have a high level of efficacy.
- Be detoxified by different biochemical pathways.
- Have similar persistence in the soil (if it is a residual herbicide).
- Exert negative cross-resistance.
- Synergise the activity of the other component.

No mixture is likely to have all these attributes, but the first two listed are the most important. There is a risk that mixtures will select for resistance to both components in the longer term. One practical advantage of sequences of two herbicides compared with mixtures is that a better appraisal of the efficacy of each herbicide component is possible, provided that sufficient time elapses between each application. A disadvantage with sequences is that two separate applications have to be made and it is possible that the later application will be less effective on weeds surviving the first application. If these are resistant, then the second herbicide in the sequence may increase selection for resistant individuals by killing the susceptible plants which were damaged but not killed by the first application, but allowing the larger, less affected, resistant plants to survive. This has been cited as one reason why ALS-resistant *Stellaria media* has evolved in Scotland recently (2000), despite the regular use of a sequence incorporating mecoprop, a herbicide with a different mode of action.

Herbicide Rotations

Rotation of herbicides from different chemical groups in successive years should reduce selection for resistance. This is a key element in most resistance prevention programmes. The value of this approach depends on the extent of cross-resistance, and whether multiple resistance occurs owing to the presence of several different resistance mechanisms. A practical problem can be the lack of awareness by farmers of the different groups of herbicides that exist. In Australia a scheme has been introduced in which identifying letters are included on the product label as a means of enabling farmers to distinguish products with different modes of action.

Farming Practices and Resistance

Herbicide resistance became a critical problem in Australian agriculture, after many Australian sheep farmers began to exclusively grow wheat in their pastures in the 1970s. Introduced varieties of ryegrass, while good for grazing sheep, compete intensely with wheat. Ryegrasses produce so many seeds that, if left unchecked, they can completely choke a field. Herbicides provided excellent control, while reducing soil disrupting because of less need to plough. Within little more than a decade, ryegrass and other weeds began to develop resistance. In response Australian farmers changed methods. By 1983, patches of ryegrass had become immune to Hoegrass, a family of herbicides that inhibit an enzyme called acetyl coenzyme A carboxylase.

Ryegrass populations were large, and had substantial genetic diversity, because farmers had planted many varieties. Ryegrass is cross-pollinated by wind, so genes shuffle frequently. To control its distribution farmers sprayed inexpensive Hoegrass, creating selection pressure. In addition, farmers sometimes diluted the herbicide in order to save money, which allowed some plants to survive application. When resistance appeared farmers turned to a group of herbicides that block acetolactate synthase. Once again, ryegrass in Australia evolved a kind of "cross-resistance" that allowed it to rapidly break down a variety of herbicides. Four classes of herbicides become ineffective within a few years. In 2013 only two herbicide classes, called Photosystem II and long-chain fatty acid inhibitors, were effective against ryegrass.

List of Common Herbicides

Synthetic Herbicides

- 2,4-D is a broadleaf herbicide in the phenoxy group used in turf and no-till field crop production. Now, it is mainly used in a blend with other herbicides to allow lower rates of herbicides to be used; it is the most widely used herbicide in the world, and third most commonly used in the United States. It is an example of synthetic auxin (plant hormone).
- Aminopyralid is a broadleaf herbicide in the pyridine group, used to control weeds on grassland, such as docks, thistles and nettles. It is notorious for its ability to persist in compost.
- Atrazine, a triazine herbicide, is used in corn and sorghum for control of broadleaf weeds and grasses. Still used because of its low cost and because it works well on a broad spectrum of weeds common in the US corn belt, atrazine is commonly used with other herbicides to reduce the overall rate of atrazine and to lower the potential for groundwater contamination; it is a photosystem II inhibitor.
- Clopyralid is a broadleaf herbicide in the pyridine group, used mainly in turf, rangeland, and for control of noxious thistles. Notorious for its ability to persist in compost, it is another example of synthetic auxin.
- Dicamba, a postemergent broadleaf herbicide with some soil activity, is used on turf and field corn. It is another example of a synthetic auxin.
- Glufosinate ammonium, a broad-spectrum contact herbicide, is used to control weeds after the crop emerges or for total vegetation control on land not used for cultivation.
- Fluazifop (Fuselade Forte), a post emergence, foliar absorbed, translocated grass-selective herbicide with little residual action. It is used on a very wide range of broad leaved crops for control of annual and perennial grasses.
- Fluroxypyr, a systemic, selective herbicide, is used for the control of broadleaved weeds in small grain cereals, maize, pastures, rangeland and turf. It is a synthetic auxin. In cereal growing, fluroxypyr's key importance is control of cleavers, Galium aparine. Other key broadleaf weeds are also controlled.
- Glyphosate, a systemic nonselective herbicide, is used in no-till burndown and for weed control in crops genetically modified to resist its effects. It is an example of an EPSPs inhibitor.
- Imazapyr a nonselective herbicide, is used for the control of a broad range of weeds, including terrestrial annual and perennial grasses and broadleaf herbs, woody species, and riparian and emergent aquatic species.

- Imazapic, a selective herbicide for both the pre- and postemergent control of some annual and perennial grasses and some broadleaf weeds, kills plants by inhibiting the production of branched chain amino acids (valine, leucine, and isoleucine), which are necessary for protein synthesis and cell growth.
- Imazamox, an imidazolinone manufactured by BASF for postemergence application that is an acetolactate synthase (ALS) inhibitor. Sold under trade names Raptor, Beyond, and Clearcast.
- Linuron is a nonselective herbicide used in the control of grasses and broadleaf weeds. It works by inhibiting photosynthesis.
- MCPA (2-methyl-4-chlorophenoxyacetic acid) is a phenoxy herbicide selective for broadleaf plants and widely used in cereals and pasture.
- Metolachlor is a pre-emergent herbicide widely used for control of annual grasses in corn and sorghum; it has displaced some of the atrazine in these uses.
- Paraquat is a nonselective contact herbicide used for no-till burndown and in aerial destruction of marijuana and coca plantings. It is more acutely toxic to people than any other herbicide in widespread commercial use.
- Pendimethalin, a pre-emergent herbicide, is widely used to control annual grasses and some broad-leaf weeds in a wide range of crops, including corn, soybeans, wheat, cotton, many tree and vine crops, and many turfgrass species.
- Picloram, a pyridine herbicide, mainly is used to control unwanted trees in pastures and edges of fields. It is another synthetic auxin.
- Sodium chlorate (disused/banned in some countries), a nonselective herbicide, is considered phytotoxic to all green plant parts. It can also kill through root absorption.
- Triclopyr, a systemic, foliar herbicide in the pyridine group, is used to control broadleaf weeds while leaving grasses and conifers unaffected.
- Several sulfonylureas, including Flazasulfuron and Metsulfuron-methyl, which act as ALS inhibitors and in some cases are taken up from the soil via the roots.

Organic Herbicides

Recently, the term “organic” has come to imply products used in organic farming. Under this definition, an organic herbicide is one that can be used in a farming enterprise that has been classified as organic. Depending on the application, they may be less effective than synthetic herbicides and are generally used along with cultural and mechanical weed control practices.

Homemade organic herbicides include:

- Corn gluten meal (CGM) is a natural pre-emergence weed control used in turfgrass, which reduces germination of many broadleaf and grass weeds.
- Vinegar is effective for 5–20% solutions of acetic acid, with higher concentrations most effective, but it mainly destroys surface growth, so respraying to treat regrowth is needed. Resistant plants generally succumb when weakened by respraying.
- Steam has been applied commercially, but is now considered uneconomical and inadequate. It controls surface growth but not underground growth and so respraying to treat regrowth of perennials is needed.
- Flame is considered more effective than steam, but suffers from the same difficulties.
- D-limonene (citrus oil) is a natural degreasing agent that strips the waxy skin or cuticle from weeds, causing dehydration and ultimately death.
- Saltwater or salt applied in appropriate strengths to the rootzone will kill most plants.
- Monocerin produced by certain fungi will kill certain weeds such as Johnson grass.

Other Herbicides

- 2,4,5-Trichlorophenoxyacetic acid (2,4,5-T) was a widely used broadleaf herbicide until being phased out starting in the late 1970s. While 2,4,5-T itself is of only moderate toxicity, the manufacturing process for 2,4,5-T contaminates this chemical with trace amounts of 2,3,7,8-tetrachlorodibenzo-p-dioxin (TCDD). TCDD is extremely toxic to humans. With proper temperature control during production of 2,4,5-T, TCDD levels can be held to about .005 ppm. Before the TCDD risk was well understood, early production facilities lacked proper temperature controls. Individual batches tested later were found to have as much as 60 ppm of TCDD. 2,4,5-T was withdrawn from use in the USA in 1983, at a time of heightened public sensitivity about chemical hazards in the environment. Public concern about dioxins was high, and production and use of other (non-herbicide) chemicals potentially containing TCDD contamination was also withdrawn. These included pentachlorophenol (a wood preservative) and PCBs (mainly used as stabilizing agents in transformer oil). Some feel that the 2,4,5-T withdrawal was not based on sound science. 2,4,5-T has since largely been replaced by dicamba and triclopyr.
- Agent Orange was a herbicide blend used by the British military during the Malayan Emergency and the U.S. military during the Vietnam War between January 1965 and April 1970 as a defoliant. It was a 50/50 mixture of the *n*-butyl esters of 2,4,5-T and 2,4-D. Because of TCDD contamination in the 2,4,5-T component, it has been blamed for serious illnesses in many people who were exposed to it.

- Diesel, and other heavy oil derivatives, are known to be informally used at times, but are usually banned for this purpose.

Molluscicide

Molluscicides are pesticides which kill mollusks, an animal phylum of tens of thousands of invertebrate creatures. Mollusks include octopi and squid, as well as snails and slugs, which are usually targeted by molluscicides.

Slugs and snails are well-known by farmers and gardeners as some of the most annoying and destructive pests. They can destroy the leaves and fruit of a large variety of plants, old and young, according to the University of California Integrated Pest Management Program.

"Snails and slugs feed on a variety of living plants and on decaying plant matter," UC IPM explained. "They chew irregular holes with smooth edges in leaves and flowers and can clip succulent plant parts. They also can chew fruit and young plant bark. Because they prefer succulent foliage or flowers, they primarily are pests of seedlings and herbaceous plants, but they also are serious pests of ripening fruits that are close to the ground such as strawberries, artichokes, and tomatoes. They also will feed on foliage and fruit of some trees; citrus are especially susceptible to damage."

Luckily, these pests are easily noticeable by the silvery trails of dried mucus left behind by their foot. When farmers and gardeners spot the telltale signs and damage left behind by snails and slugs, it's time to look to molluscicides for help.

As the U.S. Army Corps of Engineers explained, various types of molluscicides typically have similar modes of action. Molluscicides, sometimes referred to as bait, are touched or ingested by the slug or snail, then the chemicals manipulate the water balance of the animal. By affecting the amount of water in the mollusk's body, the molluscicides put the slugs or snails on the path to organ failure and inhibit healthy body functions.

There are two forms of chemical molluscicides: oxidizing and non-oxidizing. The non-oxidizing are among the most popular because they're more cost efficient. Although expensive in small quantities, non-oxidizing molluscicides are more effective in these small doses.

The Army Corps of Engineers outlined some of the most common non-oxidizing molluscicides including:

- Quanternary and polyquaternary ammonium compounds.
- Aromatic hydrocarbons.

- Endothall as the mono salt.
- Metals like copper sulfate and their salts.
- Niclosamide.

Oxidizing compounds that can be used to combat snail and slug infestations include chlorine, chlorine dioxide, chloramines, ozone, bromine, hydrogen peroxide and potassium permanganate.

One of the most commonly used chemicals in molluscicides is metaldehyde. Used in Antimillace, Ariotox, Deadline, Halizan, Limatox, Namekil and Slug-Tox, this chemical has been popular since the late 1960s for attracting and killing slugs and snails.

Other uses for Molluscicides

Aside from slugs and snails in the garden or on a farm, mollusks can also harm commercial and industrial business in bodies of water, Marrone Bio innovations explained. Mollusks can clog pipes, damage equipment and costs businesses significant amounts of money in lost efficiency.

Often people turn to molluscicides to get rid of these pests and keep business profitable. However, there are a variety of environmental factors to consider to avoid wide-spread pollution in the waterway.

Acaricide

Acaricide is a kind of pesticide, which is specially used to prevent and control the harmful acarids in the arachnids. Such as Nisuo Lang, bromide mite, dicofol, dicofol, four mites, triazole tin, single formamidine, thiophene ketone, quinacridone, mite net, speed mites, etc. Most acaricide with contact action or systemic effect in different developmental stages of mites often show some selectivity, some acaricides on adult mites, young mites and eggs are effective, some can only kill mites and eggs invalid, some can only kill eggs, known as the ovicidal agent.Appropriate acaricides should be selected based on the type of mite and the period of control. Such as tetradifon has a strong contact effect on the mite eggs and larvae, but the effect of adult mites poor. A variety of organophosphorus insecticides, such as omethoate, monocrotophos, isocarbophos, also has acaricidal effect, and can be used in rotation with acaricides to prevent resistant mites or delay the production of drug resistance. After the experiment, Acaricides can be used together with pesticides and fungicides to combat a variety of pests and mites.

In 1944, the United States StaufferChemical Co. developed the one-chlorimicarb sulfone, which only acaricidal effect. Since then, the rapid development of special acaricide, new varieties continue to emerge. The main acaricides appearing in each period: in the late 1940s, there were dinocap, neotran, chlorfenson, aramite. In the 1950s, there were dimite, benzilan, three dicofol, three chlorocide sulfone. In the 1960s, there were binapacryl, dinobuton, propargite, morestan, bromopropylate, chlorfensulphide, Chlordimeform, cyhexatin. In the 1970s, there were citrazon, fenbutatin oxide, azocyclotin, amitraz, phenproxide. In the 1980s, there were acaridyl ether, thiophene, clofentezine, nisolang. In order to solve the problem of the rapid development of resistance to insecticides, people are constantly studying and developing a new type of insecticide, which can prevent and cure the insecticide resistant mites. In order to reduce the pollution of pesticides on the environment, it is a direction to develop high efficient acaricide. Nisorang's successful development is a good example. It is effective in the prevention and control of mites concentration of only 30-50 mg / L, valid for 60 to 70 days.

There are more varieties of acaricides, mainly organic chlorine, organic sulfur, nitrobenzene, organic tin, amidine, heterocyclic. Other types include sulfur, dinitrophenols, diphenyl alcohols, diphenoxymethanes, diphenylsulfonates, diphenylsulfonates, sulfinate esters, diphenyls Quinolines, sulfonamides, hydroxamates, amidines, benzimidazoles, phenylhydrazones, organotins, antibiotics, etc.

Acaricides are mostly low toxicity, relatively safe for humans and animals. The resistance rate of mites to the insecticide is relatively fast. There is usually no cross-resistance between different types of acaricides. In order to control drug-resistant mites and delay the development of resistance to mites, we should pay more attention to rotation use of different types of acaricide or mixed use. The existing acaricides have no absorption and conduction, spraying must be uniform and thoughtful. Some acaricides kill egg activity is high, do not kill into mites, and some are opposite, so the use of such two types of drugs often can improve the activity of killing mites. Special acaricides multi no killing effect.

Piscicide

The normal function of fish toxicants (piscicides) in fisheries management is to destroy unwanted fish. Attempts to destroy one species completely while protecting other species have generally been unsuccessful, although the use of TFM (trifluoromethyl nitrophenol) to control lampreys, Antimycin to control scalefish in rearing ponds for channel catfish, or Squoxin (1,1'-methylenedi-2-napthol) to destroy squawfish are examples of poisons which have proved reasonably successful in this respect.

Fewer investigations have been made into the use of chemicals to sample fish populations with the intention of obtaining information which may otherwise be difficult to acquire. Lennon et al. (1970) reviewed about 40 chemicals which have been used as toxicants for fish or other aquatic life, but only two such toxicants, Rotenone and Antimycin, seem to have been used for fish sampling. The first being probably the oldest and most commonly used of all piscicides, and the second one of the most recently-introduced types. This review attempts to summarize the main characteristics of both the piscicides and the techniques used for sampling.

Properties of Piscicides

All toxic chemicals used in fisheries management, even anaesthetics, are capable of causing death of fish if a sufficient concentration is used. No piscicide, however, is limited in its effects to those on fish, and many chemicals are lethal, at concentrations used on fish, to planktonic and benthic invertebrates, reptiles and amphibians, and occasionally toxic (but not necessarily lethal) to birds or mammals. The time of year at which the chemicals are used must be selected carefully. Some amphibians in Europe are in danger of extinction as the result of environmental pressures, while the destruction of fish which are a source of food for fish-eating birds could be serious at times when the young are being reared. Where the water being treated is destined for human consumption, both the potential toxicity and organoleptic properties must be considered.

Hall (1975) reported that sodium cyanide, cresol, copper sulphate and toxaphene have been used in waters of the Tennessee Valley Authority (TVA), U.S.A., in addition to rotenone, the most commonly-used fish toxicant. Antonioni and Baumann (1975) described the use of Antimycin A as a management and sampling tool. Of the others, sodium cyanide is unacceptable in water supply reservoirs in view of its mammalian toxicity, cresol is likely to cause tainting and toxaphene is undesirably persistent. Copper sulphate, at concentrations likely to kill fish, is very toxic to venegation. Meyer et al. (1976) reviewed the system introduced in the United States in 1972 for the registration of chemicals used in fisheries management. Rotenone and Antimycin were approved for use as piscicides, but fish contaminated by them are not acceptable for human consumption, primarily because no suitable methods are available for residue analysis.

Rotenone

Rotenone, the active constituent of derris root, is generally accepted as the chemical for fisheries management purposes with the fewest disadvantages in use. It is lethal, at concentrations used against fish, to many species of zooplankton and to some other invertebrates, but the effects on these elements of the fauna are not generally long-lasting. It is not dangerous to birds or mammals except at very high concentrations, but it, or the other chemicals present in commercial formulations, will taint water used for drinking purposes at concentration lethal to fish.

The stability or persistence of rotenone is dependent on pH, temperature, oxygen concentration and the presence of suspended matter, among other factors, but in practice, when used in lakes, the toxic effects usually last only for a period of a few days at summer temperatures. In winter, however, at near-zero temperatures the water of treated lakes can remain toxic for weeks (Holden, unpublished). Rotenone can be destroyed rapidly by treatment of the water with potassium permanganate, an effective technique which adds significantly to operational costs. An appropriate concentration of permanganate is 1 mg/l, but 5 mg/l may itself be toxic. Alternatively, activated carbon can be used in small but unpolluted streams (Engstrom-Heg, 1974). In dense weed beds, distribution of poison is difficult, and rotenone is rapidly destroyed.

As with all chemicals, sensitivity to rotenone varies appreciably between species, and between fry, juveniles and adults of one species, and is affected by temperature and certain chemical factors such as pH. Meadows (1973) found that to ensure a complete kill in a few hours it would be necessary to use 1 mg/l of a 5 percent rotenone formulation for common perch and at least 8 mg/l for crucian carp. The toxicity increased with temperature, but hardness had little influence. The concentration most commonly used of formulations containing 5 percent rotenone appears to be 0.5 or 1.0 mg/l (0.025–0.050 mg/l rotenone), which is capable of killing most fish within an hour or two at temperatures of 15°C or more, but at much higher temperatures death may result in a few minutes. At temperatures below 10°C, death may take several hours. As young fish are usually affected first, larger predatory fish may gorge on the smaller individuals before they themselves are affected.

Antimycin

Antonioni and Baumann (1975) described the merits of Antimycin as a fish poison, one advantage being that, unlike rotenone, fish do not avoid it. This could be an important advantage in some situations, when an escape route was available, or where natural springs or other areas of low concentration of toxin were present in which the fish could survive. The toxicity of Antimycin varies considerably with species, the LC_{100} value at 12°C being 0.8 microgrammes/1 for fish such as rainbow and brown trout, and yellow perch, but 120 microgrammes/1 for black bullheads. The latter concentration is of the same order as that usually used for rotenone. Marking (1975) found that the toxicity of Antimycin decreased gradually from pH 6.5 to pH 8.5, and abruptly above pH 8.5, which may limit

Chapter 5

Weed and Pest Control: Methods and Techniques

There are various methods and techniques of weed and pest control. Some of these are mechanical weed control, bush regeneration, soil steam sterilization, soil solarization, etc. This chapter discusses in detail these methods and techniques, related to weed and pest control.

Mechanical Weed Control

Mechanical weed control is any physical activity that inhibits unwanted plant growth. Mechanical, or manual, weed control techniques manage weed populations through physical methods that remove, injure, kill, or make the growing conditions unfavorable. Some of these methods cause direct damage to the weeds through complete removal or causing a lethal injury. Other techniques may alter the growing environment by eliminating light, increasing the temperature of the soil, or depriving the plant of carbon dioxide or oxygen. Mechanical control techniques can be either selective or non-selective. A selective method has very little impact on non-target plants where as a non-selective method affects the entire area that is being treated. If mechanical control methods are applied at the optimal time and intensity, some weed species may be controlled or even eradicated.

Mechanical Control Methods

Weed Pulling

Pulling methods uproot and remove the weed from the soil. Weed pulling can be used to control some shrubs, tree saplings, and herbaceous plants. Annuals and tap-rooted weeds tend to be very susceptible to pulling. Many species are able to re-sprout from root segments that are left in the soil. Therefore, the effectiveness of this method is dependent on the removal of as much of the root system as possible. Well established perennial weeds are much less effectively controlled because of the difficulty of removing all of the root system and perennating plant parts. Small herbaceous weeds may be pulled by hand but larger plants may require the use of puller tools like the Weed Wrench or the Root Talon. This technique has a little to no impact on neighboring, non-target plants and has a minimal effect on the growing environment. However, pulling is labor-intensive and time consuming making it a more suitable method to use for small weed infestations.

Mowing

Mowing methods cut or shred the above ground of the weed and can prevent and reduce seed populations as well as restrict the growth of weeds. Mowing can be a very successful control method for many annual weeds. Mowing is the most effective when it is performed before the weeds are able to set seed because it can reduce the number of flower stalks and prevent the spread of more seed. However, the biology of the weed must be considered before mowing. Some weed species may sprout with increased vigor after being mowed. Also, some species are able to re-sprout from stem or root segments that are left behind after mowing. Brush cutting and weed eating are also mowing techniques that reduce the biomass of the weeds. Repeatedly removing biomass causes reduced vigor in many weed species. This method is usually used in combination with other control methods such as burning or herbicide treatments.

Mulching

Mulch is a layer of material that is spread on the ground. Compared with some other methods of weed control, mulch is relatively simple and inexpensive. Mulching smothers the weeds by excluding light and providing a physical barrier to impede their emergence. Mulching is successful with most annual weeds, however, some perennial weeds are not affected. Mulches may be organic or synthetic. Organic mulches consist of plant by products such as: pine straw, wood chips, green waste, compost, leaves, and grass clippings. Synthetic mulches, also known as ground cover fabric, can be made from materials like polyethylene, polypropylene, or polyester. The effectiveness of mulching is mostly dependent on the material used. Organic and synthetic mulches may be used in combination with each other to increase the amount of weeds controlled.

Tillage

Tillage, also known as cultivation, is the turning over of the soil. This method is more often used in agricultural crops. Tillage can be performed on a small scale with tools such as small, hand pushed rotary tillers or on a large scale with tractor mounted plows. Tillage is able to control weeds because when the soil is overturned, the vegetative parts of the plants are damaged and the root systems are exposed causing desiccation. Generally, the younger the weed is, the more readily it can be controlled with tillage. To control mature perennial weeds, repeated tillage is necessary. By continually destroying new growth and damaging the root system, the weed's food stores are depleted until it can no longer re-sprout. Also, when the soil is overturned, the soil seed bank is disrupted which can cause dormant weed seeds to germinate in the absence of the previous competitors. These new weeds can also be controlled by continued tillage until the soil seed bank is depleted.

Soil Solarization

Soil solarization is a simple method of weed control that is accomplished by covering the soil with a layer of clear or black plastic. The plastic that is covering the ground traps heat energy from the sun and raises the temperature of the soil. Many weed seeds and vegetative propagules are not able to withstand the temperatures and are killed. For this method to be most effective, it should be implemented during the summer months and the soil should be moist. Also, cool season weeds are more susceptible to soil solarization than are warm season weeds. Using black plastic as a cover excludes light which can help to control plants that are growing whereas clear plastic has been shown to produce higher soil temperatures.

Fire

Burning and flaming can be economical and practical methods of weed control if used carefully. For most plants, fire causes the cell walls to rupture when they reach a temperature of 45 °C to 55 °C. Burning is commonly used to control weeds in forests, ditches, and roadsides. Burning can be used to remove accumulated vegetation by destroying the dry, matured plant matter as well as killing the green new growth. Buried weed seeds and plant propagules may also be destroyed during burning, however, dry seeds are much less susceptible to the increased temperature. Flaming is used on a smaller scale and includes the use of a propane torch with a fan tip. Flaming may be used to control weeds along fences and paved areas or places where the soil may be too wet to hoe, dig, or till. Flaming is most effective on young weeds that are less than two inches tall but repeated treatments may control tougher perennial weeds.

Flooding

Flooding is a method of control that requires the area being treated to be saturated at a depth of 15 to 30 cm for a period of 3 to 8 weeks. The saturation of the soil reduces the availability of oxygen to the plant roots thereby killing the weed. This method has been shown to be highly effective in controlling establish perennial weeds and may also suppress annual weeds by reducing the weed seed populations.

Effects of Mechanical Control on the Environment

Mechanical methods of weed control cause physical changes in the immediate environment that may cause positive or negative effects. The suppression of the targeted weeds will open niches in the environment and may also stimulate the growth of other weeds by decreasing their competition and making their environment more favorable. If the niches are not filled by a desirable plant, they will eventually be taken over by another weed. These weed control methods also effect the structure of the soil. The use of mulches can help decrease erosion, decrease water evaporation from the soil, as well as improve the soil structure by increasing the amount of organic matter. Tillage practices can help decrease compaction and aerate the soil. On the other hand, tillage has also been shown to decrease soil moisture, increase soil erosion and runoff, as well as decrease soil microbial populations. Solarization can cause changes in the biological, physical, and chemical properties of the soil. This can cause the soil to be an unfavorable environment for native species which may be beneficial or harmful.

Bush Regeneration

Bush regeneration, a form of natural area restoration, is the term used in Australia for the ecological restoration of remnant vegetation areas, such as through the minimisation of negative disturbances, both exogenous such as exotic weeds and endogenous such as erosion. It may also attempt to recreate conditions of pre-European arrival, for example by simulating endogenous disturbances such as fire. Bush regeneration attempts to protect and enhance the floral biodiversity in an area by providing conditions conducive to the recruitment and survival of native plants. Bushcare's Major Day Out is an Australian national day of community participation in the care of bushland. In 2012 nearly 100 bushcare sites participated in this annual event. For more information go to www.bushcaresmajordayout.org.

History

Bradley Method

In the early 1960s Joan and Eileen Bradley developed a series of weed control techniques through a process of trial and error. Their work was the beginning of minimal disturbance bush regeneration in New South Wales. The Bradley method urges a naturalistic approach by encouraging the native vegetation to self-reestablish. The Bradleys used their method to successfully clear weeds from a 16 hectares (40 acres) reserve in Ashton Park, part of Sydney Harbour National Park, NSW. The process demonstrated that, following a period of consecutive 'follow up' treatments of diminishing time requirement, subsequent maintenance was needed only once or twice a year, mainly in vulnerable spots such as creek banks, roadsides, and clearings, to be maintained weed-free.

The aim of their work was to clear small niches adjacent to healthy native vegetation such that the each area will regenerate from in-situ soil seed banks or be re-colonised and stabilized by the regeneration of native plants, replacing an area previously occupied by weeds. The Bradley method follows three main principles;

- Secure the best areas first.
- Minimise disturbance to the natural conditions (e.g. minimise soil disturbance and off-target damage).

- Don't overclear, let the regenerative ability of the bush set the pace of clearance (Bradley 1988).

The priority securing of the best quality vegetation aids in preserving areas of top biodiversity which provide regeneration potential to expand these areas and reclaim areas as bushland.

Modern Practice

The adoption of minimal disturbance bush regeneration increased in the decades that followed the work of the Bradleys. Their principles have guided bushcare programs in Australia, although the inclusion of herbicide in modern bush regeneration is a notable deviation from the ideals of the Bradley sisters. In addition, rather than 'minimal disturbance', a more favoured and ecologically sound trend since the 1990s has been towards more 'appropriate disturbance' as many Australian plant communities require some level of perturbation to trigger germination from long-buried seed banks. This has led to a range of additional disturbance-based techniques (such as burns and soil disturbance) being included in the regenerator's 'tool kit' in dry forest and grassland areas. Field experience has found that, even in rainforest areas, a resilience to disturbance is evident, enabling regenerators to clear weed in a fairly extensive manner to trigger rainforest recovery. This is borne out by a thriving rainforest regeneration industry in northern NSW Australia, modelled on the pioneering work of John Stockard at Wingham Brush (Stockard 1991, Stockard 1999). The rule of thumb in all cases is to constrain clearing to that area that matches the project's follow up resources.

The increased awareness and consideration of Australia's biodiversity by citizens has incrementally increased pressure on local governments to adopt conservation programs for remnant vegetation on council land. Most peri urban councils now have some involvement in bush regeneration, either through planning, land management, volunteer support or through employment of bush regeneration practitioners. In NSW the level of coordination of bush regeneration programs through local governments is high, although in some other areas at present a lack of coordination is a serious concern in bush regeneration on public land, with only 40% of councils liaising with other councils. In such areas there may be a need for strategic management at a regional scale through Natural Resource Management Boards or non government organisations such as Trees For Life, which are involved in bushcare programs across wider areas.

Purposes

The aim of bush regeneration, also known as 'natural area restoration', is to restore and maintain ecosystem health by facilitating the natural regeneration of indigenous flora, this is usually achieved by selectively reducing the competitive interaction with invasive species, or mitigation of negative influences such as weeds or erosion.

Invasive plant species are often the greatest threat to remnant vegetation, and therefore bush regeneration is closely associated with weed abatement activities. Weed management as one aim of bush regeneration, is used to increase native plant recruitment. The management of factors such as fire and herbivory can be just as important, depending on the ecosystem being restored. In recent years research and on-ground management has begun to recognize the importance of ecosystem processes rather ecosystem composition and structure and research into other ways of facilitating native plant recruitment is increasing.

Technique

The original Bradley method of bush regeneration focuses on facilitating native plant recruitment from the seedbank, rather than planting seedlings or sowing seeds, as follows:

"Weeding a little at a time from the bush towards the weeds takes the pressure off the natives under favourable conditions. Native seeds and spores are ready in the ground and the natural environment favours plants that have evolved in it. The balance is tipped back towards regeneration. Keep it that way, by always working where the strongest area of bush meets the weakest weeds"

Currently the term 'bush regeneration' includes activities other than weed removal, such as replanting and introducing species into an area where soil, water, or fire regimes have shifted the type of plant appropriate to the area (e.g. a stormwater drain).

Weed species can be important habitat for native fauna (e.g. blackberry is important habitat for wrens and the southern brown bandicoot) and this should be taken into consideration with bush regeneration, for example by not clearing invasive species until adequate habitat alternatives have been established nearby with native vegetation.

Problems can occur when insufficient follow-up is conducted as the success of bush regeneration is dependent on allowing the native vegetation to regenerate in the area where weeds have been removed.

Soil Steam Sterilization

Soil steam sterilization (soil steaming) is a farming technique that sterilizes soil with steam in open fields or greenhouses. Pests of plant cultures such as weeds, bacteria, fungi and viruses are killed through induced hot steam which causes their cell structure to physically degenerate. Biologically, the method is considered a partial disinfection. Important heat-resistant, spore-forming bacteria survive and revitalize the soil after cooling down. Soil fatigue can be cured through the release of nutritive substances blocked within the soil. Steaming leads to a better starting position, quicker growth and strengthened resistance against plant disease and pests. Today, the application of hot steam is considered the best and most effective way to disinfect sick soil, potting soil and compost. It is being used as an alternative to bromomethane, whose production and use was curtailed by the Montreal Protocol. "Steam effectively kills pathogens by heating the soil to levels that cause protein coagulation or enzyme inactivation."

Benefits of Soil Steaming

Soil sterilization provides secure and quick relief of soils from substances and organisms harmful to plants such as:

- Bacteria
- Viruses
- Fungi
- Nematodes and
- Other Pests

Further positive effects are:

- All weed and weed seeds are killed.
- Significant increase of crop yields.
- Relief from soil fatigue through activation of chemical – biological reactions.
- Blocked nutritive substances in the soil are tapped and made available for plants.
- Alternative to Methyl Bromide and other critical chemicals in agriculture.

Steaming with Superheated Steam

Through modern steaming methods with superheated steam at 180–200 °C, an optimal soil disinfection can be achieved. Soil only absorbs a small amount of humidity. Micro organisms become active once the soil has cooled down. This creates an optimal environment for instant tillage with seedlings and seeds. Additionally the method of integrated steaming can promote a target-oriented resettlement of steamed soil with beneficial organisms. In the process, the soil is first freed from all organisms and then revitalized and microbiologically buffered through the injection of a soil activator based on compost which contains a natural mixture of favorable microorganisms (e.g. Bacillus subtilis, etc.).

Different types of such steam application are also available in practice, including substrate steaming and surface steaming.

Surface Steaming

Several methods for surface steaming are in use amongst which are: area sheet steaming, the steaming hood, the steaming harrow, the steaming plough and vacuum steaming with drainage pipes or mobile pipe systems.

In order to pick the most suitable steaming method, certain factors have to be considered such as soil structure, plant culture and area performance. At present, more advanced methods are being developed, such as sandwich steaming or partially integrated sandwich steaming in order to minimize energy and cost as much as possible.

Sheet Steaming

Sheet steaming with a MSD/moeschle steam boiler (left side).

Large area sheet steaming in greenhouses using a steam injector.

Surface steaming with special sheets (sheet steaming) is a method which has been established for decades in order to steam large areas reaching from 15 to 400 m^2 in one step. If properly applied, sheet steaming is simple and highly economic. The usage of heat resistant, non-decomposing insulation fleece saves up to 50% energy, reduces the steaming time significantly and improves penetration. Single working step areas up to 400 m^2 can be steamed in 4–5 hours down to 25–30 cm depth / 90°C. The usage of heat resistant and non-decomposing synthetic insulation fleece, 5 mm thick, 500 gr / m^2, can reduce steaming time by about 30%. Through a steam injector or a perforated pipe, steam is injected underneath the sheet after it has been laid out and weighted with sand sacks.

The area performance in one working step depends on the capacity of the steam generator (e.g. steam boiler):

Steam capacity kg/h:	100	250	300	400	550	800	1000	1350	2000
Area m^2:	15-20	30-50	50-65	60-90	80-120	130-180	180-220	220-270	300-400

The steaming time depends on soil structure as well as outside temperature and amounts to 1-1.5 hours per 10 cm steaming depth. Hereby the soil reaches a temperature of about 85°C. Milling for soil loosening is not recommended since soil structure may become too fine which reduces its penetrability for steam. The usage of spading machines is ideal for soil loosening. The best results can be achieved if the soil is cloddy at greater depth and granulated at lesser depth.

In practice, working with at least two sheets simultaneously has proven to be highly effective. While one sheet is used for steaming the other one is prepared for steam injection, therefore unnecessary steaming recesses are avoided.

Depth Steaming with Vacuum

Steaming with vacuum which is induced through a mobile or fixed installed pipe system in the depth of the area to be steamed, is the method that reaches the best penetration. Despite high capital cost, the fixed installation of drainage systems is reasonable for intensively used areas since steaming depths of up to 80 cm can be achieved.

In contrast to fixed installed drainage systems, pipes in mobile suction systems are on the surface. A central suction pipeline consisting of zinc-coated, fast-coupling pipes are connected in a regular spacing of 1.50 m and the ends of the hoses are pushed into the soil to the desired depth with a special tool.

The steaming area is covered with a special steaming sheet and weighted all around as with sheet steaming. The steam is injected underneath the sheet through an injector and protection tunnel. While with short areas up to 30 m length steam is frontally injected, with longer areas steam is induced in the middle of the beet using a T-connection branching out to both sides. As soon as the sheet is inflated to approximately 1m by the steam pressure, the suction turbine is switched on. First, the air in the soil is removed via the suction hoses. A vacuum is formed and the steam is pulled downward.

During the final phase, when the required steaming depth has been reached, the ventilator runs non-stop and surplus steam is blown out. To ensure that this surplus steam is not lost, it is fed back under the sheet.

As with all other steaming systems, a post-steaming period of approximately 20–30 minutes is required. Steaming time is approximately 1 hour per 10 cm steaming depth. The steam requirement is approximately 7–8 kg/m^2.

The most important requirement, as with all steaming systems, is that the soil is well loosened before steaming, to ensure optimal penetration.

Negative Pressure Technique

"Negative Pressure technique generates appropriate soil temperature at a 60 cm depth and complete control of nematodes, fungi and weeds is achieved. In this technique, the steam is introduced under the steaming sheath and forced to enter the soil profile by a negative pressure. The negative pressure is created by a fan that sucks the air out of the soil through buried perforated polypropylene pipes. This system requires a permanent installation of perforated pipes into the soil, at a depth of at least 60 cm to be protected from plough."

Steaming with Hoods

Half automatic steaming hood with three wings in greenhouse.

A steaming hood is a mobile device consisting of corrosion-resistant materials such as aluminum, which is put down onto the area to be steamed. In contrast to sheet steaming, cost-intensive working steps such as laying out and weighting the sheets don't occur, however the area steamed per working step is smaller in accordance to the size of the hood.

Outdoors, a hood is positioned either manually or via tractor with a special pre-stressed 4 point suspension arm. Steaming time amounts to 30 min for a penetration down to 25 cm depth. Hereby a temperature of 90°C can be reached. In large stable glasshouses, the hoods are attached to tracks. They are lifted and moved by pneumatic cylinders. Small and medium-sized hoods up to 12m^2 are lifted manually using a tipping lever or moved electrically with special winches.

Combined Surface and Depth Injection of Steam (Sandwich Steaming)

Sandwich steaming machine model Sterilter constructed by Ferrari Costruzioni Meccaniche equipped with MSD/moeschle steam boiler.

Sandwich steaming, which was developed in a project among DEIAFA, University of Turin (Italy, www.deiafa.unito.it) and Ferrari Costruzioni Meccaniche, represents a combination of depth and surface steaming, offers an efficient method to induce hot steam into the soil. The steam is simultaneously pushed into the soil from the surface and from the depth. For this purpose, the area, which must be equipped with a deep steaming injection system, is covered with a steaming hood. The steam enters the soil from the top and the bottom at the same time. Sheets are not suitable, since a high pressure up to 30 mm water column arises underneath the cover.

Sandwich steaming offers several advantages. On the one hand, application of energy can be increased to up to 120 kg steam per m^2/h. In comparison to other steaming methods up to 30% energy savings can be achieved and the usage of fuel (e.g. heating oil) accordingly decreases. The increased application of energy leads to a quick heating of the soil which reduces the loss of heat. On the other hand, only half of the regular steaming time is needed.

Comparison of sandwich steaming with other steam injection methods relating to steam output and energy demand(*):

Steaming method	max. steam output	energy demand (*)
Sheet steaming	6 kg/m^2h	about 100 kg $steam/m^3$
Depth steaming (Sheet + vacuum)	14 kg/m^2h	about 120 kg $steam/m^3$
Hood steaming (Alu)	30 kg/m^2h	about 80 kg $steam/m^3$
Hood steaming (Steel)	50 kg/m^2h	about 75 kg $steam/m^3$
Sandwich steaming	120 kg/m^2h	about 60 kg $steam/m^3$

(*) in soil max 30% moisture

Clearly, Sandwich steaming reaches the highest steam output at the lowest energy demand.

Partially Integrated Sandwich Steaming

The partial integrated Sandwich steaming is an advanced combined method for steaming merely the areas which shall be planted and purposely leaving out those areas which shall not be used. In order to avoid risk of re-infection of steamed areas with pest from unsteamed areas, beneficial organisms can directly be injected into the hygenized soil via a soil activator (e.g. special compost). The partial sandwich steaming unlocks further potential savings in the steaming process.

Container / Stack Steaming

Stack steaming is used when thermically treating compost and substrates such as turf. Depending on the amount, the material to be steamed is piled up to 70 cm height in steaming boxes or in small dump trailers. Steam is evenly injected via manifolds. For huge amounts, steaming containers and soil boxes are used which are equipped with suction systems to improve steaming results. Midget amounts can be steamed in special small steaming devices.

The amount of soil steamed should be tuned in a way that steaming time amounts to at most 1.5 h in order to avoid large quantities of condensed water in the bottom layers of the soil.

Steam Output kg/h:	100	250	300	400	550	800	1000	1350	2000
m^3/h about:	1.0-1.5	2.5-3.0	3.0-3.5	4.0-5.0	5.5-7.0	8.0-10.0	10.0-13.0	14.0-18.0	20.0-25.0

In light substrates, such as turf, the performance per hour is significantly higher.

History

Ancient civilizations in India and Egypt used steam, generated through the targeted usage of incident solar radiation on watered top soil, to sanitize and revive their arable land.

Modern soil steam sterilization was first discovered in 1888 (by Frank in Germany) and was first commercially used in the United States (by Rudd) in 1893 (Baker 1962). Since then, a wide variety of steam machines have been built to disinfest both commercial greenhouse and nursery field soils (Grossman and Liebman 1995). In the 1950s, for example, steam sterilization technologies expanded from disinfestation of potting soil and greenhouse mixes to commercial production of steam rakes and tractor-drawn steam blades for fumigating small acres of cut flowers and other high-value field crops (Langedijk 1959). Today, even more effective steam technologies are being developed.

Application of Hot Steam

- In horticulture as well as nurseries for sterilization of substrates and top soil.
- In agriculture for sterilization and treatment of food waste for pig fattening and heating of molasses.
- In mushroom cultivation for pasteurization of growing rooms, sterilization of top soil and combined application as heating.
- In wineries as combination boiler for sterilization and cleaning of storage tanks, tempering of mash and for warm water generation.

Soil Solarization

Soil solarization is an environmentally friendly method of using solar power for controlling pests such as soilborne plant pathogens including fungi, bacteria, nematodes, and insect and mite pests along with weed seed and seedlings in the soil by mulching the soil and covering it with tarp, usually with a transparent polyethylene cover, to trap solar energy. It may also describe methods of decontaminating soil using sunlight or solar power. This energy causes physical, chemical, and biological changes in the soil.

Soil Disinfestation

Soil solarization (referred to as solar heating of the soil in early publications) is a relatively new soil disinfestation method, first described in extensive scientific detail by Katan et al. in 1976, presenting the results of a series of studies performed under field conditions, initiated in 1973, for controlling soilborne pathogens and weeds, mostly as a pre-planting soil treatment. Soil is mulched and then covered with transparent polyethylene during the hot season, thereby heating it and killing the pests.

Soil Decontamination

A 2008 study used a solar cell to generate an electric field for electrokinetic (EK) remediation of cadmium-contaminated soil. The solar cell could drive the electromigration of cadmium in contaminated soil, and the removal efficiency that was achieved by the solar cell was comparable with that achieved by conventional power supply.

In Korea, various remediation methods of soil slurry and groundwater contaminated with benzene at a polluted gas station site were evaluated, including a solar-driven, photocatalyzed reactor system along with various advanced oxidation processes (AOP). The most synergistic remediation method incorporated a solar light process with TiO_2 slurry and H_2O_2 system, achieving 98% benzene degradation, a substantial increase in the removal of benzene.

History

Attempts were made to use solar energy for controlling disease agents in soil and in plant material already in the ancient civilization of India. In 1939, Groashevoy, who used the term "solar energy for sand disinfection," controlled Thielaviopsis basicola upon heating the sand by exposure to direct sunlight.

Soil solarization is the third approach for soil disinfestation; the two other main approaches, soil steaming and fumigation; were developed at the end of the 19th century. The idea of solarization was based on observations by extension workers and farmers in the hot Jordan Valley, who noticed the intensive heating of the polyethylene-mulched soil. The involvement of biological control mechanisms in pathogen control and the possible implications were indicated in the first publication, noticing the very long effect of the treatment. In 1977, American scientists from the University of California at Davis reported the control of Verticillium in a cotton field, based on studies started in 1976, thus denoting, for the first time, the possible wide applicability of this method.

The use of polyethylene for soil solarization differs in principle from its traditional agricultural use. With solarization, soil is mulched during the hottest months (rather than the coldest, as in conventional plasticulture which is aimed at protecting the crop) in order to increase the maximal temperatures in an attempt to achieve lethal heat levels.

In the first 10 years following the influential 1976 publication, soil solarization was investigated in at least 24 countries and has been now been applied in more than 50, mostly in the hot regions, although there were some important exceptions. Studies have demonstrated effectiveness of solarization with various crops, including vegetables, field crops, ornamentals and fruit trees, against many pathogens, weeds and a soil arthropod. Those pathogens and weeds which are not controlled by solarization were also detected. The biological, chemical and physical changes that take in solarized soil during and after the solarization have been investigated, as well as the interaction of solarization with other methods of control. Long-term effects including biological control and increased growth response were verified in various climatic regions and soils, demonstrating the general applicability of solarization. Computerized simulation models have been developed to guide researchers and growers whether the ambient conditions of their locality are suitable for solarization.

Studies of the improvement of solarization by integrating it with other methods or by solarizing in closed glasshouses, or studies concerning commercial application by developing mulching machines were also carried out.

The use of solarization in existing orchards (e.g. controlling Verticillium in pistachio plantations) is an important deviation from the standard preplanting method and was reported as early as 1979.

Stale Seed Bed

A false or stale seed bed is a useful weed control technique which involves creating a seedbed some weeks before seed is due to be sown. Preparation of such seedbed makes sure that any weed seeds that have been disturbed and brought to the soil surface during cultivation will thus have a chance to germinate, and can then be hoed off or eliminated with the use of a flame weeder before sowing of the actual crop is carried out.

The technique can be utilized in early spring, when the weather is still too cold for proper seed germination. Several passes are made with a rototiller or plow, then weed seeds are allowed to germinate as weather permits.

By tilling, the farmer increases the chance of weed seed germination by the same method as one would for favorable vegetable/crops: the fine soil allows weed seed to grow rapidly by allowing the seed to open and the roots to spread easier than in compacted soil. Deep tilling will also bring dormant seed to the surface for germination. Some species of plant are known for seeds that can lay deeply buried in the soil for years before favorable conditions allow germination.

Timing is important, however. Weed seeds must be tilled/howed or otherwise destroyed before they themselves can create new seeds. By destroying them early, the farmer eliminates most of that season's annual weeds. Turning the dead weeds back into the soil also increases soil nutrient content, although this difference is slight.

In many cases, several tillings are done, perhaps every two weeks beginning in very early spring. This allows more and more weed seeds to germinate only to be killed off later. This eliminates more weeds, but care must be used to not delay planting of a desirable crop later than the crop needs for a successful season's growth.

After several years, most, if not all, weeds can be eliminated from the seed bank in the soil. In some cases the effect can be noticed in the same year the process is first carried out.

If the weed patch is vigorous enough, the farmer may mow the patch first before tilling. This allows for easier/quicker decomposition in the soil when the plants are turned under.

Some farmers are noted for applying a light and inexpensive fertilizer mix to the soil to cause even more weed seeds to germinate. Although the merits of this may seem a trifle at first, it can pay dividends in the long run by eliminating seeds earlier that otherwise would have sprouted in later years. This too is open to debate.

Aerial Application

Aerial application, or what was formerly referred to as crop dusting, involves spraying crops with crop protection products from an agricultural aircraft. Planting certain types of seed are also included in aerial application. The specific spreading of fertilizer is also known as *aerial topdressing* in some countries.

Agricultural aircraft are highly specialized, purpose-built aircraft. Today's agricultural aircraft are often powered by turbine engines of up to 1500 hp and can carry as much as 800 gallons of crop protection product. Helicopters are sometimes used, and some aircraft serve double duty as water bombers in areas prone to wildfires.(These aircraft are referred to as S.E.A.T. "single engine air tankers").

History

Aerial Seed Sowing 1906

The first known aerial application of agricultural materials was by John Chaytor, who in 1906 spread seed over a swamped valley floor in Wairoa, New Zealand, using a hot air balloon with mobile tethers. Aerial sowing of seed still continues to this day with cover crop applications and rice planting.

Crop Dusting 1921

Lt. Macready (right) and McCook Field engineer E. Dormoy (left) in front of the 1st crop duster airplane (August 3, 1921).

The first known use of a heavier-than-air machine to disperse products occurred on 3 August 1921. Crop dusting was developed under the joint efforts of the U.S. Agriculture Department, and the U.S. Army Signal Corps's research station at McCook Field in Dayton, Ohio. Under the direction of McCook engineer Etienne Dormoy, a United States Army Air Service Curtiss JN4 Jenny piloted by John A. Macready was modified at McCook Field to spread lead arsenate to kill catalpa sphinx caterpillars at a Catalapa farm near Troy, Ohio in the United States. The first test was considered highly successful. The first commercial operations were begun in 1924, by Huff-Daland Crop Dusting, which was co-founded by McCook Field test pilot Lt. Harold R. Harris. Use of insecticide and fungicide for crop dusting slowly spread in the Americas and to a lesser extent other nations in the 1930s. The name 'crop dusting' originated here, as actual dust was spread across the crops. Today, aerial applicators use liquid crop protection products in very small doses.

Top Dressing 1939–1946

Aerial topdressing, the spread of fertilizers such as superphosphate, was developed in New Zealand in the 1940s by members of the Ministry of Public Works and RNZAF, led by Alan Pritchard and Doug Campbell - unofficial experiments by individuals within the government led to funded research. Initially fertilizer and seed were dropped together (1939), using a window mounted chute on a Miles Whitney Straight, but by the end of the 1940s different mixtures of fertilizer were being distributed from hoppers installed in war surplus Grumman Avengers and C-47 Skytrains, as well as some privately operated de Havilland Tiger Moths in New Zealand, and the practise was being adopted experimentally in Australia and the United Kingdom.

Purpose Built Aircraft

In 1951, Leland Snow begins designing the first aircraft specifically built for aerial application, the S-1.

In 1957, The Grumman G-164 Ag-Cat is the first aircraft designed by a major company for ag aviation.

Water Bombing 1952

Aerial firefighting, or water bombing, was tested experimentally by Art Seller's Skyways air services in Canada in 1952 (dropping a mix of water, fertilizer and seed), and established in California in the mid-1950s.

Night Aerial Application 1973–Present

Aerial application at night is mostly liquid spray and is conducted in the Southwest U.S. deserts. The increased heat, scheduling conflicts with farm workers in the fields and honeybee activity reduced the effectiveness of spraying in daytime. In high temperature areas, the insects would travel down in plants in daytime and return to the top at night. The aircraft — both fixed wing, autogyros and helicopters — were equipped with lights, usually three sets: Work lights were high power and aimed or adjustable from the cockpit; wire lights were angled down for taxiing and wire or obstruction illumination; and turn lights were only turned on in the direction of the turn to allow safe operation on moonless nights where angle of entry or exit needed to be illuminated. Some aircraft were equipped with an elongated metal wing called a spreader, with inbuilt channels to direct the flow of dust such as sulfur, used on melons as a pesticide and soil amendment. Very little pesticide dust was used day or night in comparison to spray, because of the difficulty in drift control. Workers on the ground, called "flaggers", would use flashlights aimed at the aircraft to mark the swaths on the ground; later, GPS units replaced the flaggers due to new laws restricting use of human flaggers with some pesticides. GPS systems also provide precise guidance for the applicator.

Agricultural chemicals have also kept pace with advancements in technology, and have been influential in the growth of the agricultural aviation industry. In the 1930s Aerial Applicators arrived in the northern states to war against insect and disease pests which threatened fruit and vegetable crops. After World War II, the industry expanded into the western states where the development of new chemicals made possible the control of weeds and insects in cereal grain crops. Some of these new chemicals proved very useful in controlling various insects that carried diseases dangerous to humans. Countries that previously had no control over malaria and river blindness were provided with chemicals which helped save hundreds of thousands of lives and reduced the suffering of millions. All during the 1950s, crop production continued to rise and disease declined as a result of chemical controls.

Aerial application accounts for just under 20% of all applied crop protection products on commercial farms. The industry also provides firefighting and public health application services According to a 2012 NAAA survey, the five most common aerially treated crops are: corn, wheat/barley, soybeans, pastures/rangelands and alfalfa, but aerial application is used on many more crops grown in the U.S.

Approximately 1,350 aerial application businesses are in the U.S. and 1,430 non-operator pilots. 94% of aerial application business owners (operators) are also pilots. Aerial application businesses are located in 44 states – all but Connecticut, Hawaii, Nevada, Rhode Island, Vermont and West Virginia.

Today's ag aircraft use sophisticated precision application equipment such as: GPS (global positioning systems), GIS (geographical information systems), Aircraft Integrated Meteorological Measurement System (AIMMS), real time meteorological systems, flow control valves for variable-rate applications, single-boom shutoff valves and smokers to identify wind speed and direction.

According to the U.S. Bureau of Labor Statistics, in 2005 U.S. cropduster pilots earned an average annual wage of $63,210.

Unmanned Aerial Application

A Yamaha R-MAX, a UAV commonly used for aerial application in Japan.

Beginning in the late 1990s, unmanned aerial vehicles are also being used for agricultural spraying. This phenomenon started in Japan and South Korea, where mountainous terrain and relatively small family-owned farms required lower-cost and higher precision spraying. As of 2014, the use of UAV crop dusters, such as the Yamaha R-MAX, is being expanded to the United States for use in spraying of vineyards.

Irrigation

Irrigation is the method in which water is supplied to plants at regular intervals for agriculture. It is used to assist in the growing of agricultural crops, maintenance of landscapes, and revegetation of disturbed soils in dry areas and during periods of inadequate rainfall. Additionally, irrigation also has a few other uses in crop production, which include protecting plants against frost, suppressing weed growth in grain fields and preventing soil consolidation. In contrast, agriculture that relies only on direct rainfall is referred to as rain-fed or dry land farming.

Irrigation systems are also used for dust suppression, disposal of sewage, and in mining. Irrigation is often studied together with drainage, which is the natural or artificial removal of surface and sub-surface water from a given area.

Irrigation has been a central feature of agriculture for over 5,000 years and is the product of many cultures. Historically, it was the basis for economies and societies across the globe, from Asia to the Southwestern United States.

History

Animal-powered irrigation, Upper Egypt, ca. 1846.

An example of an irrigation system common on the Indian subcontinent. Artistic impression on the banks of Dal Lake, Kashmir, India.

Inside a karez tunnel at Turpan, Sinkiang.

Archaeological investigation has identified as evidence of irrigation where the natural rainfall was insufficient to support crops for rainfed agriculture.

Perennial irrigation was practiced in the Mesopotamian plain whereby crops were regularly watered throughout the growing season by coaxing water through a matrix of small channels formed in the field.

irrigation in Tamil Nadu (India).

Ancient Egyptians practiced *Basin irrigation* using the flooding of the Nile to inundate land plots which had been surrounded by dykes. The flood water was held until the fertile sediment had settled before the surplus was returned to the watercourse. There is evidence of the ancient Egyptian pharaoh Amenemhet III in the twelfth dynasty (about 1800 BCE) using the natural lake of the Faiyum Oasis as a reservoir to store surpluses of water for use during the dry seasons, the lake swelled annually from flooding of the Nile.

The Ancient Nubians developed a form of irrigation by using a waterwheel-like device called a *sakia*. Irrigation began in Nubia some time between the third and second millennium BCE. It largely depended upon the flood waters that would flow through the Nile River and other rivers in what is now the Sudan.

In sub-Saharan Africa irrigation reached the Niger River region cultures and civilizations by the first or second millennium BCE and was based on wet season flooding and water harvesting.*Terrace irrigation* is evidenced in pre-Columbian America, early Syria, India, and China. In the Zana Valley of the Andes Mountains in Peru, archaeologists found remains of three irrigation canals radiocarbon dated from the 4th millennium BCE, the 3rd millennium BCE and the 9th century CE. These canals are the earliest record of irrigation in the New World. Traces of a canal possibly dating from the 5th millennium BCE were found under the 4th millennium canal. Sophisticated irrigation and storage systems were developed by the Indus Valley Civilization in present-day Pakistan and North India, including the reservoirs at Girnar in 3000 BCE and an early canal irrigation system from circa 2600 BCE. Large scale agriculture was practiced and an extensive network of canals was used for the purpose of irrigation.

Ancient Persia (modern day Iran) as far back as the 6th millennium BCE, where barley was grown in areas where the natural rainfall was insufficient to support such a crop. The Qanats, developed in ancient Persia in about 800 BCE, are among the oldest known irrigation methods still in use today. They are now found in Asia, the Middle East and North Africa. The system comprises a network of vertical wells and gently sloping tunnels driven into the sides of cliffs and steep hills to tap groundwater. The noria, a water wheel with clay pots around the rim powered by the flow of the stream (or by animals where the water source was still), was first brought into use at about this time, by Roman settlers in North Africa. By 150 BCE the pots were fitted with valves to allow smoother filling as they were forced into the water.

The irrigation works of ancient Sri Lanka, the earliest dating from about 300 BCE, in the reign of King Pandukabhaya and under continuous development for the next thousand years, were one of the most complex irrigation systems of the ancient world. In addition to underground canals, the Sinhalese were the first to build completely artificial reservoirs to store water. Due to their engineering superiority in this sector, they were often called 'masters of irrigation'. Most of these irrigation systems still exist undamaged up to now, in Anuradhapura and Polonnaruwa, because of the advanced and precise engineering. The system was extensively restored and further extended during the reign of King Parakrama Bahu (1153–1186 CE).

China

The oldest known hydraulic engineers of China were Sunshu Ao (6th century BCE) of the Spring and Autumn Period and Ximen Bao (5th century BCE) of the Warring States period, both of whom worked on large irrigation projects. In the Sichuan region belonging to the State of Qin of ancient China, the Dujiangyan Irrigation System was built in 256 BCE to irrigate an enormous area of farmland that today still supplies water. By the 2nd century AD, during the Han Dynasty, the Chinese also used chain pumps that lifted water from lower elevation to higher elevation. These were powered by manual foot pedal, hydraulic waterwheels, or rotating mechanical wheels pulled by oxen. The water was used for public works of providing water for urban residential quarters and palace gardens, but mostly for irrigation of farmland canals and channels in the fields.

Korea

In 15th century Korea, the world's first rain gauge, *uryanggye* (Korean:□□□), was invented in 1441. The inventor was Jang Yeong-sil, a Korean engineer of the Joseon Dynasty, under the active direction of the king, Sejong the Great. It was installed in irrigation tanks as part of a nationwide system to measure and collect rainfall for agricultural applications. With this instrument, planners and farmers could make better use of the information gathered in the survey.

North America

In North America, the Hohokam were the only culture to rely on irrigation canals to water their crops, and their irrigation systems supported the largest population in the Southwest by AD 1300. The Hohokam constructed an assortment of simple canals combined with weirs in their various agricultural pursuits. Between the 7th and 14th centuries, they also built and maintained extensive irrigation networks along the lower Salt and middle Gila rivers that rivaled the complexity of those used in the ancient Near East, Egypt, and China. These were constructed using relatively simple excavation tools, without the benefit of advanced engineering technologies, and achieved drops of a few feet per mile, balancing erosion and siltation. The Hohokam cultivated varieties of cotton, tobacco, maize, beans and squash, as well as harvested an assortment of wild plants. Late in the Hohokam Chronological Sequence, they also used extensive dry-farming systems, primarily to grow agave for food and fiber. Their reliance on agricultural strategies based on canal irrigation, vital in their less than hospitable desert environment and arid climate, provided the basis for the aggregation of rural populations into stable urban centers.

Present Extent

In the mid-20th century, the advent of diesel and electric motors led to systems that could pump groundwater out of major aquifers faster than drainage basins could refill them. This can lead to permanent loss of aquifer capacity, decreased water quality, ground subsidence, and other problems. The future of food production in such areas as the North China Plain, the Punjab, and the Great Plains of the US is threatened by this phenomenon.

Irrigation ditch in Montour County, Pennsylvania, off Strawberry Ridge Road.

At the global scale, 2,788,000 km^2 (689 million acres) of fertile land was equipped with irrigation infrastructure around the year 2000. About 68% of the area equipped for irrigation is located in Asia, 17% in the Americas, 9% in Europe, 5% in Africa and 1% in Oceania. The largest contiguous areas of high irrigation density are found:

- In Northern India and Pakistan along the Ganges and Indus rivers.
- In the Hai He, Huang He and Yangtze basins in China.
- Along the Nile river in Egypt and Sudan.
- In the Mississippi-Missouri river basin and in parts of California.

Smaller irrigation areas are spread across almost all populated parts of the world.

Only eight years later in 2008, the scale of irrigated land increased to an estimated total of 3,245,566 km^2 (802 million acres), which is nearly the size of India.

Types of Irrigation

Basin flood irrigation of wheat.

Irrigation of land in Punjab, Pakistan.

Various types of irrigation techniques differ in how the water obtained from the source is distributed within the field. In general, the goal is to supply the entire field uniformly with water, so that each plant has the amount of water it needs, neither too much nor too little.

Surface Irrigation

In *surface* (*furrow*, *flood*, or *level basin*) irrigation systems, water moves across the surface of agricultural lands, in order to wet it and infiltrate into the soil. Surface irrigation can be subdivided into furrow, *borderstrip or basin irrigation*. It is often called *flood irrigation* when the irrigation results in flooding or near flooding of the cultivated land. Historically, this has been the most common method of irrigating agricultural land and still is in most parts of the world.

Where water levels from the irrigation source permit, the levels are controlled by dikes, usually plugged by soil. This is often seen in terraced rice fields (rice paddies), where the method is used to flood or control the level of water in each distinct field. In some cases, the water is pumped, or lifted by human or animal power to the level of the land. The field water efficiency of surface irrigation is typically lower than other forms of irrigation but has the potential for efficiencies in the range of 70% - 90% under appropriate management.

Localized Irrigation

Impact sprinkler head.

Localized irrigation is a system where water is distributed under low pressure through a piped network, in a pre-determined pattern, and applied as a small discharge to each plant or adjacent to it. Drip irrigation, spray or micro-sprinkler irrigation and bubbler irrigation belong to this category of irrigation methods.

Subsurface Textile Irrigation

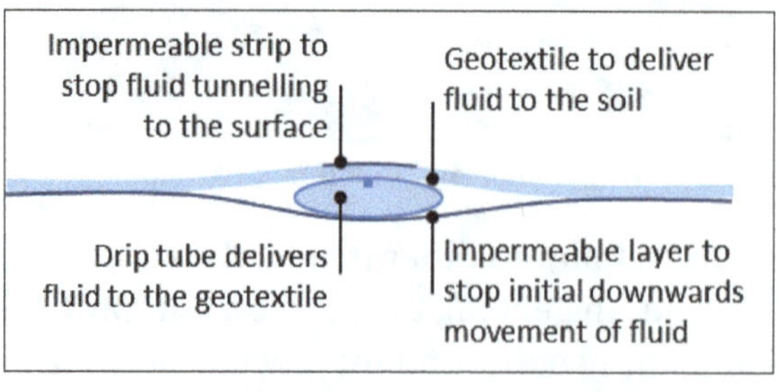

Diagram showing the structure of an example SSTI installation.

Subsurface Textile Irrigation (SSTI) is a technology designed specifically for subsurface irrigation in all soil textures from desert sands to heavy clays. A typical subsurface textile irrigation system has an impermeable base layer (usually polyethylene or polypropylene), a drip line running along that base, a layer of geotextile on top of the drip line and, finally, a narrow impermeable layer on top of the geotextile. Unlike standard drip irrigation, the spacing of emitters in the drip pipe is not critical as the geotextile moves the water along the fabric up to 2 m from the dripper.

Drip Irrigation

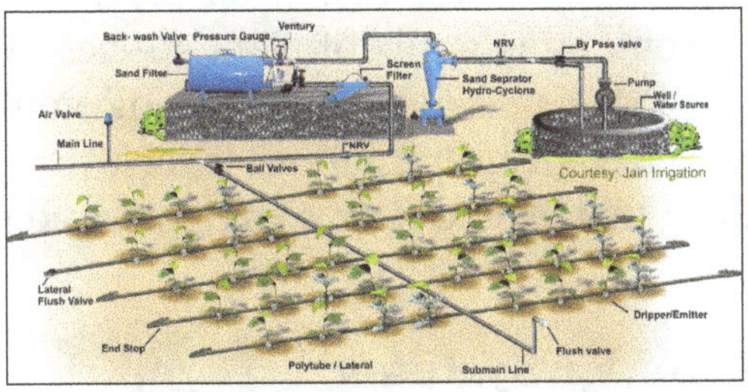

Drip irrigation layout and its parts.

Drip irrigation – a dripper in action.

Grapes in Petrolina, only made possible in this semi arid area by drip irrigation.

Drip (or micro) irrigation, also known as trickle irrigation, functions as its name suggests. In this system water falls drop by drop just at the position of roots. Water is delivered at or near the root zone of plants, drop by drop. This method can be the most water-efficient method of irrigation, if managed properly, since evaporation and runoff are minimized. The field water efficiency of drip irrigation is typically in the range of 80 to 90 percent when managed correctly.

In modern agriculture, drip irrigation is often combined with plastic mulch, further reducing evaporation, and is also the means of delivery of fertilizer. The process is known as *fertigation*.

Deep percolation, where water moves below the root zone, can occur if a drip system is operated for too long or if the delivery rate is too high. Drip irrigation methods range from very high-tech and computerized to low-tech and labor-intensive. Lower water pressures are usually needed than for most other types of systems, with the exception of low energy center pivot systems and surface irrigation systems, and the system can be designed for uniformity throughout a field or for precise water delivery to individual plants in a landscape containing a mix of plant species. Although it is difficult to regulate pressure on steep slopes, pressure compensating emitters are available, so the field does not have to be level. High-tech solutions involve precisely calibrated emitters located along lines of tubing that extend from a computerized set of valves.

Irrigation Using Sprinkler Systems

Sprinkler irrigation of blueberries in Plainville, New York, United States.

A traveling sprinkler at Millets Farm Centre, Oxfordshire, United Kingdom.

In *sprinkler* or overhead irrigation, water is piped to one or more central locations within the field and distributed by overhead high-pressure sprinklers or guns. A system utilizing sprinklers, sprays, or guns mounted overhead on permanently installed risers is often referred to as a *solid-set* irrigation system. Higher pressure sprinklers that rotate are called *rotors* an are driven by a ball drive, gear drive, or impact mechanism.

Rotors can be designed to rotate in a full or partial circle. Guns are similar to rotors, except that they generally operate at very high pressures of 40 to 130 lbf/in^2 (275 to 900 kPa) and flows of 50 to 1200 US gal/min (3 to 76 L/s), usually with nozzle diameters in the range of 0.5 to 1.9 inches (10 to 50 mm). Guns are used not only for irrigation, but also for industrial applications such as dust suppression and logging.

Sprinklers can also be mounted on moving platforms connected to the water source by a hose. Automatically moving wheeled systems known as *traveling sprinklers* may irrigate areas such as small farms, sports fields, parks, pastures, and cemeteries unattended. Most of these utilize a length of polyethylene tubing wound on a steel drum. As the tubing is wound on the drum powered by the irrigation water or a small gas engine, the sprinkler is pulled across the field. When the sprinkler arrives back at the reel the system shuts off. This type of system is known to most people as a "waterreel" traveling irrigation sprinkler and they are used extensively for dust suppression, irrigation, and land application of waste water.

Other travelers use a flat rubber hose that is dragged along behind while the sprinkler platform is pulled by a cable. These cable-type travelers are definitely old technology and their use is limited in today's modern irrigation projects.

Irrigation using Center Pivot

A small center pivot system from beginning to end.

Center pivot irrigation.

Center pivot irrigation is a form of sprinkler irrigation consisting of several segments of pipe (usually galvanized steel or aluminium) joined together and supported by trusses, mounted on wheeled towers with sprinklers positioned along its length. The system moves in a circular pattern and is fed with water from the pivot point at the center of the arc. These systems are found and used in all parts of the world and allow irrigation of all types of terrain. Newer systems have drop sprinkler heads as shown in the image that follows.

Most center pivot systems now have drops hanging from a u-shaped pipe attached at the top of the pipe with sprinkler head that are positioned a few feet (at most) above the crop, thus limiting evaporative losses. Drops can also be used with drag hoses or bubblers that deposit the water directly on the ground between crops. Crops are often planted in a circle to conform to the center pivot. This type of system is known as LEPA (Low Energy Precision Application). Originally, most center pivots were water powered. These were replaced by hydraulic systems (*T-L Irrigation*) and electric motor driven systems (Reinke, Valley, Zimmatic). Many modern pivots feature GPS devices.

Irrigation by Lateral Move (Side Roll, Wheel Line, Wheelmove)

A *series of pipes, each with a wheel* of about 1.5 m diameter permanently affixed to its midpoint, and sprinklers along its length, are coupled together. Water is supplied at one end using a large hose. After sufficient irrigation has been applied to one strip of the field, the hose is removed, the water drained from the system, and the assembly rolled either by hand or with a purpose-built mechanism, so that the sprinklers are moved to a different position across the field. The hose is reconnected. The process is repeated in a pattern until the whole field has been irrigated.

This system is less expensive to install than a center pivot, but much more labor-intensive to operate - it does not travel automatically across the field: it applies water in a stationary strip, must be drained, and then rolled to a new strip. Most systems use 4 or 5-inch (130 mm) diameter aluminum pipe. The pipe doubles both as water transport and as an axle for rotating all the wheels. A drive system (often found near the centre of the wheel line) rotates the clamped-together pipe sections as a single axle, rolling the whole wheel line. Manual adjustment of individual wheel positions may be necessary if the system becomes misaligned.

Wheel line systems are limited in the amount of water they can carry, and limited in the height of crops that can be irrigated. One useful feature of a lateral move system is that it consists of sections that can be easily disconnected, adapting to field shape as the line is moved. They are most often used for small, rectilinear, or oddly-shaped fields, hilly or mountainous regions, or in regions where labor is inexpensive.

Sub-Irrigation

Subirrigation has been used for many years in field crops in areas with high water tables. It is a method of artificially raising the water table to allow the soil to be moistened from below the plants' root zone. Often those systems are located on permanent grasslands in lowlands or river valleys and combined with drainage infrastructure. A system of pumping stations, canals, weirs and gates allows it to increase or decrease the water level in a network of ditches and thereby control the water table.

Sub-irrigation is also used in commercial greenhouse production, usually for potted plants. Water is delivered from below, absorbed upwards, and the excess collected for recycling. Typically, a solution of water and nutrients floods a container or flows through a trough for a short period of time, 10–20 minutes, and is then pumped back into a holding tank for reuse. Sub-irrigation in greenhouses requires fairly sophisticated, expensive equipment and management. Advantages are water and nutrient conservation, and labor-saving through lowered system maintenance and automation. It is similar in principle and action to subsurface basin irrigation.

Irrigation Automatically, Non-Electric using Buckets and Ropes

Besides the common manual watering by bucket, an automated, natural version of this also exists. Using plain polyester ropes combined with a prepared ground mixture can be used to water plants from a vessel filled with water.

The ground mixture would need to be made depending on the plant itself, yet would mostly consist of black potting soil, vermiculite and perlite. This system would (with certain crops) allow to save expenses as it does not consume any electricity and only little water (unlike sprinklers, water timers, etc.). However, it may only be used with certain crops (probably mostly larger crops that do not need a humid environment; perhaps e.g. paprikas).

Irrigation using Water Condensed from Humid Air

In countries where at night, humid air sweeps the countryside.Water can be obtained from the humid air by condensation onto cold surfaces. This is for example practiced in the vineyards at Lanzarote using stones to condense water or with various fog collectors based on canvas or foil sheets.

In-ground Irrigation

Most commercial and residential irrigation systems are "in ground" systems, which means that everything is buried in the ground. With the pipes, sprinklers, emitters (drippers), and irrigation valves being hidden, it makes for a cleaner, more presentable landscape without garden hoses or other items having to be moved around manually. This does, however, create some drawbacks in the maintenance of a completely buried system.

Most irrigation systems are divided into zones. A zone is a single irrigation valve and one or a group of drippers or sprinklers that are connected by pipes or tubes. Irrigation systems are divided into zones because there is usually not enough pressure and available flow to run sprinklers for an entire yard or sports field at once. Each zone has a solenoid valve on it that is controlled via wire by an irrigation controller. The irrigation controller is either a mechanical (now the "dinosaur" type) or electrical device that signals a zone to turn on at a specific time and keeps it on for a specified amount of time. "Smart Controller" is a recent term for a controller that is capable of adjusting the watering time by itself in response to current environmental conditions. The smart controller determines current conditions by means of historic weather data for the local area, a soil moisture sensor (water potential or water content), rain sensor, or in more sophisticated systems satellite feed weather station, or a combination of these.

When a zone comes on, the water flows through the lateral lines and ultimately ends up at the irrigation emitter (drip) or sprinkler heads. Many sprinklers have pipe thread inlets on the bottom of them which allows a fitting and the pipe to be attached to them. The sprinklers are usually installed with the top of the head flush with the ground surface. When the water is pressurized, the head will pop up out of the ground and water the desired area until the valve closes and shuts off that zone. Once there is no more water pressure in the lateral line, the sprinkler head will retract back into the ground. Emitters are generally laid on the soil surface or buried a few inches to reduce evaporation losses.

Water Sources

Irrigation is underway by pump-enabled extraction directly from the Gumti, seen in the background, in Comilla, Bangladesh.

Irrigation water can come from groundwater (extracted from springs or by using wells), from surface water (withdrawn from rivers, lakes or reservoirs) or from non-conventional sources like treated wastewater, desalinated water or drainage water. A special form of irrigation using surface water is spate irrigation, also called floodwater harvesting. In case of a flood (spate), water is diverted to normally dry river beds (wadis) using a network of dams, gates and channels and spread over large areas. The moisture stored in the soil will be used thereafter to grow crops. Spate irrigation areas are in particular located in semi-arid or arid, mountainous regions. While floodwater harvesting belongs to the accepted irrigation methods, rainwater harvesting is usually not considered as a form of irrigation. Rainwater harvesting is the collection of runoff water from roofs or unused land and the concentration of this.

Around 90% of wastewater produced globally remains untreated, causing widespread water pollution, especially in low-income countries. Increasingly, agriculture uses untreated wastewater as a source of irrigation water. Cities provide lucrative markets for fresh produce, so are attractive to farmers. However, because agriculture has to compete for increasingly scarce water resources with industry and municipal users, there is often no alternative for farmers but to use water polluted with urban waste, includ-ing sewage, directly to water their crops. Significant health hazards can result from using water loaded with pathogens in this way, especially if people eat raw vegetables that have been irrigated with the polluted water. The International Water Management Institute has worked in India, Pakistan, Vietnam, Ghana, Ethiopia, Mexico and other countries on various projects aimed at assessing and reducing risks of wastewater irrigation. They advocate a 'multiple-barrier' approach to wastewater use, where farmers are encouraged to adopt various risk-reducing behaviours. These include ceasing irrigation a few days before harvesting to allow pathogens to die off in the sunlight, applying water carefully so it does not contaminate leaves likely to be eaten raw, cleaning vegetables with disinfectant or allowing fecal sludge used in farming to dry before being used as a human manure. The World Health Organization has developed guidelines for safe water use.

There are numerous benefits of using recycled water for irrigation, including the low cost (when compared to other sources, particularly in an urban area), consistency of supply (regardless of season, climatic conditions and associated water restrictions), and general consistency of quality. Irrigation of recycled wastewater is also considered as a means for plant fertilization and particularly nutrient supplementation. This approach carries with it a risk of soil and water pollution through excessive wastewater application. Hence, a detailed understanding of soil water conditions is essential for effective utilization of wastewater for irrigation.

Efficiency

Modern irrigation methods are efficient enough to supply the entire field uniformly with water, so that each plant has the amount of water it needs, neither too much nor too little. Water use efficiency in the field can be determined as follows:

Young engineers restoring and developing the old Mughal irrigation system during the reign of the Mughal Emperor Bahadur Shah II.

- Field Water Efficiency (%) = (Water Transpired by Crop ÷ Water Applied to Field) x 100.

Until 1960s, the common perception was that water was an infinite resource. At that time, there were fewer than half the current number of people on the planet. People were not as wealthy as today, consumed fewer calories and ate less meat, so less water was needed to produce their food. They required a third of the volume of water we presently take from rivers. Today, the competition for water resources is much more intense. This is because there are now more than seven billion people on the planet, their consumption of water-thirsty meat and vegetables is rising, and there is increasing competition for water from industry, urbanisation and biofuel crops. To avoid a global water crisis, farmers will have to strive to increase productivity to meet growing demands for food, while industry and cities find ways to use water more efficiently.

Successful agriculture is dependent upon farmers having sufficient access to water. However, water scarcity is already a critical constraint to farming in many parts of the world. With regards to agriculture, the World Bank targets food production and water management as an increasingly global issue that is fostering a growing debate. Physical water scarcity is where there is not enough water to meet all demands, including that needed for ecosystems to function effectively. Arid regions frequently suffer from physical water scarcity. It also occurs where water seems abundant but where resources are over-committed. This can happen where there is overdevelopment of hydraulic infrastructure, usually for irrigation. Symptoms of physical water scarcity include environmental degradation and declining groundwater. Economic scarcity, meanwhile, is caused by a lack of investment in water or insufficient human capacity to satisfy the demand for water. Symptoms of economic water scarcity include a lack of infrastructure, with people often having to fetch water from rivers for domestic and agricultural uses. Some 2.8 billion people currently live in water-scarce areas.

Technical Challenges

Irrigation schemes involve solving numerous engineering and economic problems while minimizing negative environmental impact.

- Competition for surface water rights.
- Overdrafting (depletion) of underground aquifers.
- Ground subsidence (e.g. New Orleans, Louisiana).
- Underirrigation or irrigation giving only just enough water for the plant (e.g. in drip line irrigation) gives poor soil salinity control which leads to increased soil salinity with consequent buildup of toxic salts on soil surface in areas with high evaporation. This requires either leaching to remove these salts and a method of drainage to carry the salts away. When using drip lines, the leaching is best

done regularly at certain intervals (with only a slight excess of water), so that the salt is flushed back under the plant's roots.

- Overirrigation because of poor distribution uniformity or management wastes water, chemicals, and may lead to water pollution.
- Deep drainage (from over-irrigation) may result in rising water tables which in some instances will lead to problems of irrigation salinity requiring watertable control by some form of subsurface land drainage.
- Irrigation with saline or high-sodium water may damage soil structure owing to the formation of alkaline soil
- Clogging of filters: It is mostly algae that clog filters, drip installations and nozzles. UV and ultrasonic method can be used for algae control in irrigation systems.

Mechanical Pest Control

Mechanical pest control is the management and control of pests using physical means such as fences, barriers or electronic wires. It includes also weeding and change of temperature to control pests. Many farmers at the moment are trying to find sustainable ways to remove pests without harming the ecosystem.

Methods

Handpicking

The use of human hands to remove harmful insects or other toxic material is often the most common action by gardeners. It is also classified as the most direct and the quickest way to remove clearly visible pests. However, it also has equal disadvantages as it must be performed before damage to the plant has been done and before the key development of insects.

Mechanical Traps

Mechanical traps or physical attractants are used in three main ways: to efficiently trap insects, to kill them or to estimate how much many insects there are in the total landmass using sampling method. However, some traps are expensive to produce and can end up benefiting insects rather than harming them.

Differences from Integrated Pest Control

Integrated pest control refers to the use of any means to control pests once they reach unacceptable levels. Mechanical pest control is but a minor part of integrated pest control. It means only the use of physical means to control pests.

Physical Pest Control

Physical Pest Control is a method of getting rid of insects and small rodents by removing, attacking, or setting up barriers that will prevent further destruction of one's plants. These methods are used primarily for crop growing, but some methods can be applied to homes as well.

Methods

Barriers

Dog control van, Rekong Peo, Himachal Pradesh, India.

Row covers are useful for keeping insects out of one's plants, typically used for horticultural crops. They are made out of either plastic or polyester. They are made thin and light to allow plants to still absorb sunshine and water from the air.

Diatomaceous earth, made from fossilized and pulverized silica shells, can be used in order to damage the protective cuticle layer of insects that have them, such as ants. When this layer is damaged, the insects become vulnerable to drying out. Unfortunately, the effectiveness of Diatomaceous earth decreases if it is wet. Therefore, it must be used often. This method was used back in the 1930s and 1940s when farmers would run dust over their fields. This would have the very same effect as diatomaceous earth.

Traps

Fly paper or sticky boards are devices used in order to capture insects as they land upon the surface of the trap. They are covered in a substance that attracts insects, but are actually very sticky or poisonous. These traps are commonly used for flies or leafhoppers.

Trap strips are crops that are grown on fields with the intention of using them to attract insects and not have insects infest the other crops that are being grown. The insects can then be dealt with much more easily than if they were to have been spread throughout an entire field. Trap strips are very useful for dealing with the wheat stem sawfly. The sawflies will go only as far as they need to in order to plant their eggs. If solid stemmed plants are planted around the a crop field, then that's where the sawflies will go and the sawflies' larvae can't survive in the solid stem.

Fire

For farmers, fire has been a powerful technique used to destroy insect breeding grounds. It is used to burn the top of the soil in order to kill the insects that lie there. Unfortunately, this can present some drawbacks. Fire can make the soil much less effective or get rid of the insects that are beneficial to the plants. Also, there is no guarantee that it will actually solve the pest problems since there may be larvae below the surface of the soil.

Temperature Control

Placing produce inside of cold storage containers lengthens how long the produce lasts while also hindering the growth of insects inside of them. Another method to use is to heat, as it will kill the insect larvae in certain types of produce. An example would be with mangoes, where they are placed into a hot water bath in order to kill any eggs and larvae.

Large Scale Usage

On a much larger scale, physical control methods become much less effective because of the time that must be invested into it and because it is likely to be less economical. For example, taking care of a single tree is simple, but taking care of 500, like on a farm, would be impossible using physical control.

Fumigation

Fumigation is a method of pest control that completely fills an area with gaseous pesticides—or fumigants—to suffocate or poison the pests within. It is used to control pests in buildings (structural fumigation), soil, grain, and produce, and is also used during processing of goods to be imported or exported to prevent transfer of exotic organisms. This method also affects the structure itself, affecting pests that inhabit the physical structure, such as woodborers and drywood termites.

Process

Fumigation generally involves the following phases: First the area intended to be fumigated is usually covered to create a sealed environment; next the fumigant is released into the space to be fumigated; then, the space is held for a set period while the fumigant gas percolates through the space and acts on and kills any infestation in the product, next the space is ventilated so that the poisonous gases are allowed to escape from the space, and render it safe for humans to enter. If successful, the fumigated area is now safe and pest free.

Tent Fumigation (Tenting)

Structural fumigation techniques differ from building to building, but in houses a rubber tent is often placed over the entire house while the pesticides are being released into the vacant residence. This process is called tent fumigation or "tenting". The sealed tent concentrates the poisonous gases and prevents them from escaping into the neighborhood. The process can take up to a week depending on the fumigant used, which is in turn dependent on the severity of infestation and size of the building.

Chemicals

Methyl bromide was among the most widely used fumigants until its production and use was restricted by the Montreal Protocol due to its role in ozone depletion.

Widely used fumigants include:

- 1,3-dichloropropene
- Dazomet
- Chloropicrin
- DBCP, prior to 1985
- Formaldehyde
- Hydrogen cyanide
- Iodoform
- Methyl isocyanate
- Phosphine
- Sulfuryl fluoride

Safety

Fumigation is a hazardous operation. Generally it is a legal requirement that the operator who carries out the fumigation operation holds official certification to perform the fumigation as the chemicals used are toxic to most forms of life, including humans.

Post operation ventilation of the area is a critical safety aspect of fumigation. It is important to distinguish between the pack or source of the fumigant gas and the environment which has been fumigated. While the fumigant pack may be safe and spent, the space will still hold the fumigant gas until it has been ventilated.

Electronic Pest Control

Electronic pest control is the name given to the use of any of the several types of electrically powered devices designed to repel or eliminate pests, usually rodents or insects. Since these devices are not regulated under the Federal Insecticide, Fungicide and Rodenticide Act (FIFRA) in the United States, the US EPA does not require the same kind of efficacy testing that it does for chemical pesticides.

Types of Devices

There are two types of electronic pest control devices widely available: electromagnetic and ultrasonic.

Electromagnetic

Electromagnetic ("EM") pest repelling devices claim to affect the nervous system of ants, mice, spiders, and other rodents. There have been similar studies on effects of EM radiation emitted by cellphones on humans.

Ultrasonic

Ultrasonic devices operate through emitting short wavelength, high frequency sound waves that are too high in pitch to be heard by the human ear (all frequencies greater than 20,000 Hz). Humans are unable to hear sounds higher than 20 kHz due to physiological limitations of the cochlea. Some animals, such as bats, dogs, and rodents, can hear well into the ultrasonic range. Some insects, such as grasshoppers and locusts, can detect frequencies from 50,000 Hz to 100,000 Hz, and lacewings and moths can detect ultrasound as high as 240,000 Hz produced by insect-hunting bats. Contrary to popular belief, birds cannot hear ultrasonic sound. Some smartphone applications attempt to use this technology to produce high frequency sounds to repel mosquitoes and other insects, but the claims of effectiveness of these application and of ultrasonic control of mosquitoes in general has been questioned.

Insects detect sound by special hairs or sensilla located on the antennae (mosquitoes) or genitalia (cockroaches), or by more complicated tympanal organs (butterflies, grasshoppers, locusts, and moths).

Radio Wave Pest Control

The concept of radio wave (RW) or radio frequency (RF) to control the behavior of living organisms has shown promise. According to Drs. Juming Tang and Shaojin Wang at Washington State University (WSU) with colleagues at the University of California-Davis and USDA's Agricultural Research Service in Parlier, California, since RF energy generates heat through agitation of bound water molecules, it generates heat through ionic conduction and agitation of free water molecules in insects. As a result, more thermal energy is converted in insects.

RF treatments control insect pests without negatively affecting food stuffs and storage locations. RF treatments may serve as a non-chemical alternative to chemical fumigants for post-harvest pest control in commodities (such as almonds, pecans, pistachios, lentils, peas, and soybeans), reducing the long-term impact on the environment, human health, and competitiveness of agricultural industries.

Effects on Pests

Studies

"Ultrasound and Arthropod Pest Control" (2001), an extensive Kansas State University study, confirmed that ultrasonic sound devices do have both a repellent effect as well as reduces mating and reproduction of insects. However, the results were mixed, and ultrasonic sound had little or no effect on some pests. Ultrasonic devices were highly effective on crickets, while the same devices had little repellent effect on cockroaches. Additionally, the results were mixed: some devices were effective, while others had no effect depending on the test subject. The study also concluded there was no effect on ants or spiders in any of the tests. They concluded, based on the mixed results, that more research is needed to improve these devices.

A 2002 study sponsored by Genesis Laboratories, Inc. (the maker of the Pest-A-Cator/ Riddex series of electronic repellent devices) does lend some credence to the ability of electronic repellent devices to repel certain pests in controlled environments. "Preliminary study of white-footed mice behavior in the test apparatus demonstrated a significant preference for the non-activated chamber among both sexes."

In 2003, the Federal Trade Commission required Global Instruments, the maker of the Pest-A-Cator/Riddex series of electromagnetic pest control devices, to discontinue any claims for their efficacy until they are backed by credible scientific evidence. This ban continues to be in effect.

In 2009, Victor Pest obtained positive results from independent researchers which resulted in two ultrasonic devices' being granted registration by the Canadian EPA (PMRA). The results from the tests were: the device "successfully repelled the rodents from the protected area in 13 of the 17 sites. This represents a 81.3% success rate...the average number of days before rodent activity was stopped was six days".

Effects

Effects on Cockroaches

Cockroaches respond to electronic pest control devices by moving about a bit more than usual, but don't appear eager to escape from the sound waves. This includes devices that emit a uniform frequency as well as those that emit changing frequencies of ultrasound. Researchers were able to use the increased cockroach activity to good effect by increasing the rate at which they caught the roaches in sticky traps.

Effects on Mosquitoes

A 2007 review article examined 10 field studies, in which ultrasonic repellent devices had been put to the test, and concluded they "have no effect on preventing mosquito bites" and "should not be recommended or used". It goes on: "Given these findings from 10 carefully conducted studies, it would not be worthwhile to conduct further research on EMRs [electronic mosquito repellents] in preventing mosquitoes biting or in trying to prevent the acquisition of malaria".

Bart Knols, an entomologist who chairs the advisory board of the Dutch Malaria Foundation and edits the website Malaria World, claims there is "no scientific evidence whatsoever" that ultrasound repels mosquitoes.

In 2005, the British consumer magazine *Holiday* reported the results of its test of a range of mosquito deterrents. The magazine's editor Lorna Cowan described the four appliances that used a buzzer as "a shocking waste of money" which "should be removed from sale". One, the Lovebug, a ladybird-shaped gadget designed to be clipped onto a baby's cot or child's pushchair - was singled out as a particular cause for concern, because of the likelihood that parents would trust it to keep mosquitoes away, and their children would be hurt as a result. (The Lovebug is still readily available in Europe, though it was withdrawn from the US market after the Federal Trade Commission reprimanded the manufacturer Prince Lionheart.)

Effects on Rodents

Rodents adjust to the ultrasound (or any new sound) and eventually ignore it. At best, ultrasonic waves have only a partial or temporary effect on rodents. Numerous studies have rejected ultrasonic sound as a practical means of rodent control. Tests of commercial ultrasonic devices have indicated that rodents may be repelled from the immediate area of the ultrasound device for a few minutes to a few days, but they will nearly always return and resume normal activities. Other tests have shown that the degree of repellance depends on the frequency, intensity, and pre-existing condition of the rodent infestation. The intensity of such sounds must be so great that damage to humans or domestic animals would also be likely; commercial ultrasonic pest control devices do not produce sounds of such intensity.

Safety

Professor Tim Leighton at the Institute of Sound and Vibration Research], University of Southampton, U.K. produced an 83-page paper entitled "What is Ultrasound?"

(2007), in which he expressed concern about the growth in commercial products which exploit the discomforting effects of in-air ultrasound (to pests for whom it is within their audible frequency range, or to humans for whom it is not, but who can experience unpleasant subjective effects and, potentially, shifts in the hearing threshold). Leighton claims that commercial products are often advertised with cited levels which cannot be critically accepted due to lack of accepted measurement standards for ultrasound in air, and little understanding of the mechanism by which they may represent a hazard.

The UK's independent Advisory Group on Non-ionising Radiation (AGNIR) produced a 180-page report on the health effects of human exposure to ultrasound and infrasound in 2010. The UK Health Protection Agency (HPA) published their report, which recommended an exposure limit for the general public to airborne ultrasound sound pressure levels (SPL) of 70 dB (at 20 kHz), and 100 dB (at 25 kHz and above).

Biological Pest Control

Biological control is a method of controlling pests such as insects, mites, weeds and plant diseases using other organisms. It relies on predation, parasitism, herbivory, or other natural mechanisms, but typically also involves an active human management role. It can be an important component of integrated pest management (IPM) programs.

There are three basic types of biological pest control strategies: importation (sometimes called classical biological control), in which a natural enemy of a pest is introduced in the hope of achieving control; augmentation, in which locally-occurring natural enemies are bred and released to improve control; and conservation, in which measures are taken to increase natural enemies, such as by planting nectar-producing crop plants in the borders of rice fields.

Natural enemies of insect pests, also known as biological control agents, include predators, parasitoids, and pathogens. Biological control agents of plant diseases are most often referred to as antagonists. Biological control agents of weeds include seed predators, herbivores and plant pathogens.

Biological control can have side-effects on biodiversity through predation, parasitism, pathogenicity, competition, or other attacks on non-target species, especially when a species is introduced without thorough understanding of the possible consequences.

Types of Biological Pest Control

There are three basic biological pest control strategies: importation (classical biological control), augmentation and conservation.

Importation

Importation or classical biological control involves the introduction of a pest's natural enemies to a new locale where they do not occur naturally. Early instances were often unofficial and not based on research, and some introduced species became serious pests themselves.

Rodolia cardinalis, the vedalia beetle, was imported to Australia in the 19th century, successfully controlling cottony cushion scale.

To be most effective at controlling a pest, a biological control agent requires a colonizing ability which allows it to keep pace with the spatial and temporal disruption of the habitat. Control is greatest if the agent has temporal persistence, so that it can maintain its population even in the temporary absence of the target species, and if it is an opportunistic forager, enabling it to rapidly exploit a pest population.

Joseph Needham noted a Chinese text dating from 304 AD, *Records of the Plants and Trees of the Southern Regions*, by Hsi Han, which describes mandarin oranges protected by large reddish-yellow citrus ants which attack and kill insect pests of the orange trees. The citrus ant (*Oecophylla smaragdina*) was rediscovered in the 20th century, and since 1958 has been used in China to protect orange groves.

One of the earliest successes in the west was in controlling *Icerya purchasi* (cottony cushion scale) in Australia, using a predatory insect *Rodolia cardinalis* (the vedalia beetle). This success was repeated in California using the beetle and a parasitoid fly, *Cryptochaetum iceryae*.

Prickly pear cacti were introduced into Queensland, Australia as ornamental plants. They quickly spread to cover over 25 million hectares of Australia. Two control agents were used to help control the spread of the plant, the cactus moth *Cactoblastis cactorum*, and *Dactylopius* scale insects.

Damage from *Hypera postica*, the alfalfa weevil, a serious introduced pest of forage, was substantially reduced by the introduction of natural enemies. 20 years after their introduction the population of weevils in the alfalfa area treated for alfalfa weevil in the Northeastern United States remained 75 percent down.

Cactoblastis cactorum larvae feeding on *Opuntia* cacti.

The invasive species *Alternanthera philoxeroides* (alligator weed) was controlled in Florida (U.S.) by introducing alligator weed flea beetle.

Alligator weed was introduced to the United States from South America. It takes root in shallow water, interfering with navigation, irrigation, and flood control. The alligator weed flea beetle and two other biological controls were released in Florida, enabling the state to ban the use of herbicides to control alligator weed three years later. Another aquatic weed, the giant salvinia (*Salvinia molesta*) is a serious pest, covering waterways, reducing water flow and harming native species. Control with the salvinia weevil (*Cyrtobagous salviniae*) is effective in warm climates, and in Zimbabwe, a 99% control of the weed was obtained over a two-year period.

Small commercially reared parasitoidal wasps, *Trichogramma ostriniae*, provide limited and erratic control of the European corn borer (*Ostrinia nubilalis*), a serious pest. Careful formulations of the bacterium *Bacillus thuringiensis* are more effective.

The population of *Levuana iridescens*, the Levuana moth, a serious coconut pest in Fiji, was brought under control by a classical biological control program in the 1920s.

Augmentation

Augmentation involves the supplemental release of natural enemies, boosting the naturally occurring population. In inoculative release, small numbers of the control agents are released at intervals to allow them to reproduce, in the hope of setting up longer-term control, and thus keeping the pest down to a low level, constituting prevention rather than cure. In inundative release, in contrast, large numbers are released in the hope of rapidly reducing a damaging pest population, correcting a problem that has already arisen. Augmentation can be effective, but is not guaranteed to work, and relies on understanding of the situation.

Hippodamia convergens, the convergent lady beetle, is commonly sold for biological control of aphids.

An example of inoculative release occurs in greenhouse production of several crops. Periodic releases of the parasitoid, *Encarsia formosa*, are used to control greenhouse whitefly, while the predatory mite *Phytoseiulus persimilis* is used for control of the two-spotted spider mite.

The egg parasite *Trichogramma* is frequently released inundatively to control harmful moths. Similarly, *Bacillus thuringiensis* and other microbial insecticides are similarly used in large enough quantities for a rapid effect. Recommended release rates for *Trichogramma* in vegetable or field crops range from 5,000 to 200,000 per acre (1 to 50 per square metre) per week according to the level of pest infestation. Similarly, entomopathogenic nematodes are released at rates of millions and even billions per acre for control of certain soil-dwelling insect pests.

Conservation

The conservation of existing natural enemies in an environment is the third method of biological pest control. Natural enemies are already adapted to the habitat and to the target pest, and their conservation can be simple and cost-effective, as when nectar-producing crop plants are grown in the borders of rice fields. These provide nectar to support parasitoids and predators of planthopper pests and have been demonstrated to be so effective (reducing pest densities by 10- or even 100-fold) that farmers sprayed 70% less insecticides, enjoyed yields boosted by 5%, and this led to an economic advantage of 7.5%. Predators of aphids were similarly found to be present in tussock grasses by field boundary hedges in England, but they spread too slowly to reach the centres of fields. Control was improved by planting a metre-wide strip of tussock grasses in field centres, enabling aphid predators to overwinter there.

An inverted flowerpot filled with straw to attract earwigs.

Cropping systems can be modified to favor natural enemies, a practice sometimes referred to as habitat manipulation. Providing a suitable habitat, such as a shelterbelt, hedgerow, or beetle bank where beneficial insects can live and reproduce, can help ensure the survival of populations of natural enemies. Things as simple as leaving a layer of fallen leaves or mulch in place provides a suitable food source for worms and provides a shelter for insects, in turn being a food source for such beneficial mammals as hedgehogs and shrews. Compost piles and stacks of wood can provide shelter for invertebrates and small mammals. Long grass and ponds support amphibians. Not removing dead annuals and non-hardy plants in the autumn allows insects to make use of their hollow stems during winter. In California, prune trees are sometimes planted in grape vineyards to provide an improved overwintering habitat or refuge for a key grape pest parasitoid. The providing of artificial shelters in the form of wooden caskets, boxes or flowerpots is also sometimes undertaken, particularly in gardens, to make a cropped area more attractive to natural enemies. For example, earwigs are natural predators which can be encouraged in gardens by hanging upside-down flowerpots filled with straw or wood wool. Green lacewings can be encouraged by using plastic bottles with an open bottom and a roll of cardboard inside. Birdhouses enable insectivorous birds to nest; the most useful birds can be attracted by choosing an opening just large enough for the desired species.

Biological Control Agents

Predators

Predators are mainly free-living species that directly consume a large number of prey during their whole lifetime. Ladybugs, and in particular their larvae which are active between May and July in the northern hemisphere, are voracious predators of aphids, and also consume mites, scale insects and small caterpillars. The spotted lady beetle (*Coleomegilla maculata*) is also able to feed on the eggs and larvae of the Colorado potato beetle (*Leptinotarsa decemlineata*).

Lacewings are available from biocontrol dealers.

The larvae of many hoverfly species principally feed upon greenfly (aphids), one larva devouring up to 400 in its lifetime. Their effectiveness in commercial crops has not been studied.

Predatory *Polistes* wasp looking for bollworms or other caterpillars on a cotton plant.

Several species of entomopathogenic nematode are important predators of insect and other invertebrate pests. *Phasmarhabditis hermaphrodita* is a microscopic nematode that kills slugs. Its complex life cycle include a free-living, infective stage in the soil where it becomes associated with a pathogenic bacteria such as *Moraxella osloensis*. The nematode enters the slug through the posterior mantle region, thereafter feeding and reproducing inside, but it is the bacteria that kill the slug. The nematode is available commercially in Europe and is applied by watering onto moist soil.

Species used to control spider mites include the predatory mites *Phytoseiulus persimilis, Neoseilus californicus,* and *Amblyseius cucumeris*, the predatory midge *Feltiella acarisuga*, and a ladybird *Stethorus punctillum*. The bug *Orius insidiosus* has been successfully used against the two-spotted spider mite and the western flower thrips (*Frankliniella occidentalis*).

Parasitoids

Parasitoids lay their eggs on or in the body of an insect host, which is then used as a food for developing larvae. The host is ultimately killed. Most insect parasitoids are wasps or flies, and may have a very narrow host range. The most important groups are the ichneumonid wasps, which prey mainly on caterpillars of butterflies and moths; braconid wasps, which attack caterpillars and a wide range of other insects including greenfly; chalcid wasps, which parasitize eggs and larvae of greenfly, whitefly, cabbage caterpillars, and scale insects; and tachinid flies, which parasitize a wide range of insects including caterpillars, adult and larval beetles, and true bugs.

Encarsia formosa was one of the first biological control agents developed.

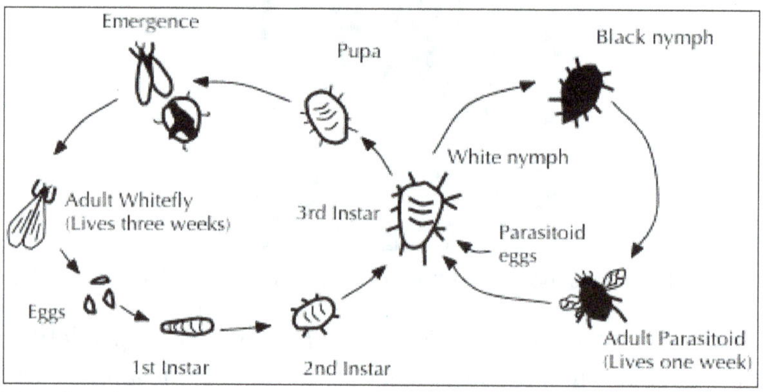

Life cycles of Greenhouse whitefly and its parasitoid wasp *Encarsia formosa*.

Encarsia formosa is a small predatory chalcid wasp which is a parasitoid of whitefly, a sap-feeding insect which can cause wilting and black sooty moulds in glasshouse vegetable and ornamental crops. It is most effective when dealing with low level infestations, giving protection over a long period of time. The wasp lays its eggs in young whitefly 'scales', turning them black as the parasite larvae pupates. *Gonatocerus ashmeadi* (Hymenoptera: Mymaridae) has been introduced to control the glassy-winged sharpshooter *Homalodisca vitripennis* (Hemipterae: Cicadellidae) in French Polynesia and has successfully controlled ~95% of the pest density.

Parasitoids are among the most widely used biological control agents. Commercially, there are two types of rearing systems: short-term daily output with high production of parasitoids per day, and long-term low daily output with a range in production of 4-1000million female parasitoids per week. Larger production facilities produce on a yearlong basis, whereas some facilities produce only seasonally. Rearing facilities are usually a significant distance from where the agents are to be used in the field, and transporting the parasitoids from the point of production to the point of use can pose problems. Shipping conditions can be too hot, and even vibrations from planes or trucks can adversely affect parasitoids.

Pathogens

Pathogenic micro-organisms include bacteria, fungi, and viruses. They kill or debilitate their host and are relatively host-specific. Various microbial insect diseases occur naturally, but may also be used as biological pesticides. When naturally occurring, these outbreaks are density-dependent in that they generally only occur as insect populations become denser.

Bacteria

Bacteria used for biological control infect insects via their digestive tracts, so they offer only limited options for controlling insects with sucking mouth parts such as aphids and scale insects. *Bacillus thuringiensis* is the most widely applied species of bacteria used for biological control, with at least four sub-species used against Lepidopteran (moth, butterfly), Coleopteran (beetle) and Dipteran (true fly) insect pests. The bacterium is available in sachets of dried spores which are mixed with water and sprayed onto vulnerable plants such as brassicas and fruit trees. *B. thuringiensis* has also been incorporated into crops, making them resistant to these pests and thus reducing the use of pesticides. The bacterium *Paenibacillus popilliae* causes milky spore disease has been found useful in the control of Japanese beetle, killing the larvae. It is very specific to its host species and is harmless to vertebrates and other invertebrates.

Fungi

Green peach aphid, a pest in its own right and a vector of plant viruses, killed by the fungus *Pandora neoaphidis* (Zygomycota: Entomophthorales) Scale bar = 0.3 mm.

Entomopathogenic fungi, which cause disease in insects, include at least 14 species that attack aphids. *Beauveria bassiana* is mass-produced and used to manage a wide variety of insect pests including whiteflies, thrips, aphids and weevils. *Lecanicillium* spp. are deployed against white flies, thrips and aphids. *Metarhizium* spp. are used against pests including beetles, locusts and other grasshoppers, Hemiptera, and spider mites. *Paecilomyces fumosoroseus* is effective against white flies, thrips and aphids; *Purpureocillium lilacinus* is used against root-knot nematodes, and 89 *Trichoderma* species against certain plant pathogens. *Trichoderma viride* has been used against Dutch elm disease, and has shown some effect in suppressing silver leaf, a disease of stone fruits caused by the pathogenic fungus *Chondrostereum purpureum*.

The fungi *Cordyceps* and *Metacordyceps* are deployed against a wide spectrum of arthropods. *Entomophaga* is effective against pests such as the green peach aphid.

Several members of Chytridiomycota and Blastocladiomycota have been explored as agents of biological control. From Chytridiomycota, *Synchytrium solstitiale* is being considered as a control agent of the yellow star thistle (*Centaurea solstitialis*) in the United States.

Viruses

Baculoviruses are specific to individual insect host species and have been shown to be useful in biological pest control. For example, the Lymantria dispar multicapsid nuclear polyhedrosis virus has been used to spray large areas of forest in North America where larvae of the gypsy moth are causing serious defoliation. The moth larvae are killed by the virus they have eaten and die, the disintegrating cadavers leaving virus particles on the foliage to infect other larvae.

A mammalian virus, the rabbit haemorrhagic disease virus has been introduced to Australia and to New Zealand to attempt to control the European rabbit populations there.

Algae

Lagenidium giganteum is a water-borne mould that parasitizes the larval stage of mosquitoes. When applied to water, the motile spores avoid unsuitable host species and search out suitable mosquito larval hosts. This alga has the advantages of a dormant phase, resistant to desiccation, with slow-release characteristics over several years. Unfortunately, it is susceptible to many chemicals used in mosquito abatement programmes.

Plants

The legume vine *Mucuna pruriens* is used in the countries of Benin and Vietnam as a biological control for problematic *Imperata cylindrica* grass. *Mucuna pruriens* is said not to be invasive outside its cultivated area. *Desmodium uncinatum* can be used in push-pull farming to stop the parasitic plant, *Striga*.

Other Methods

Combined use of Parasitoids and Pathogens

In cases of massive and severe infection of invasive pests, techniques of pest control are often used in combination. An example is the emerald ash borer, *Agrilus planipennis*, an invasive beetle from China, which has destroyed tens of millions of ash trees in its introduced range in North America. As part of the campaign against it, from 2003 American scientists and the Chinese Academy of Forestry searched for its natural enemies in the wild, leading to the discovery of several parasitoid wasps, namely *Tetrastichus planipennisi*, a gregarious larval endoparasitoid,*Oobius agrili*, a solitary, parthenogenic egg parasitoid, and *Spathius agrili*, a gregarious larval ectoparasitoid. These have been introduced and released into the United States of America as a possible biological control of the emerald ash borer. Initial results have shown promise with *Tetrastichus planipennisi* and it is now being released along with *Beauveria bassiana*, a fungal pathogen with known insecticidal properties.

Indirect Control

Pests may be controlled by biological control agents that do not prey directly upon them. For example, the Australian bush fly, *Musca vetustissima*, is a major nuisance pest in Australia, but native decomposers found in Australia are not adapted to feeding on cow dung, which is where bush flies breed. Therefore, the Australian Dung Beetle Project (1965–1985), led by Dr. George Bornemissza of the Commonwealth Scientific and Industrial Research Organisation, released forty-nine species of dung beetle, with the aim of reducing the amount of dung and therefore also the potential breeding sites of the fly.

Side-effects

Biological control can affect biodiversity through predation, parasitism, pathogenicity, competition, or other attacks on non-target species. An introduced control does not always target only the intended pest species; it can also target native species. In Hawaii during the 1940s parasitic wasps were introduced to control a lepidopteran pest and the wasps are still found there today. This may have a negative impact on the native ecosystem, however, host range and impacts need to be studied before declaring their impact on the environment.

Vertebrate animals tend to be generalist feeders, and seldom make good biological control agents; many of the classic cases of "biocontrol gone awry" involve vertebrates. For example, the cane toad (*Bufo marinus*) was intentionally introduced to Australia to control the greyback cane beetle (*Dermolepida albohirtum*), and other pests of sugar cane. 102 toads were obtained from Hawaii and bred in captivity to increase their numbers until they were released into the sugar cane fields of the tropic north in 1935. It was later discovered that the toads could not jump very high and so were unable to eat the cane beetles which stayed up on the upper stalks of the cane plants. However the toad thrived by feeding on other insects and it soon spread very rapidly; it took over native amphibian habitat and brought foreign disease to native toads and frogs, dramatically reducing their populations. Also when it is threatened or handled, the cane toad releases poison from parotoid glands on its shoulders; native Australian species such as goannas, tiger snakes, dingos and northern quolls that attempted to eat the toad were harmed or killed. However, there has been some recent evidence that native predators are adapting, both physiologically and through changing their behaviour, so in the long run, their populations may recover.

Rhinocyllus conicus, a seed-feeding weevil, was introduced to North America to control exotic musk thistle (*Carduus nutans*) and Canadian thistle (*Cirsium arvense*). However the weevil also attacks native thistles, harming such species as the endemic Platte thistle (*Cirsium neomexicanum*) by selecting larger plants (which reduced the gene pool), reducing seed production and ultimately threatening the species' survival.

The small Asian mongoose (*Herpestus javanicus*) was introduced to Hawaii in order to control the rat population. However it was diurnal and the rats emerged at night, and it preyed on the endemic birds of Hawaii, especially their eggs, more often than it ate the rats, and now both rats and mongooses threaten the birds. This introduction was undertaken without understanding the consequences of such an action. No regulations existed at the time, and more careful evaluation should prevent such releases now.

The sturdy and prolific eastern mosquitofish (*Gambusia holbrooki*) is a native of the southeastern United States and was introduced around the world in the 1930s and 40s to feed on mosquito larvae and thus combat malaria. However, it has thrived at the expense of local species, causing a decline of endemic fish and frogs through competition for food resources, as well as through eating their eggs and larvae. In Australia, the mosquitofish is the subject of discussion as to how best to control it; in 1989 it was said that "biological population control is well beyond present capabilities", and this remains the position.

Grower Education

A potential obstacle to the adoption of biological pest control measures is growers sticking to the familiar use of pesticides. It has been claimed that many of the pests that are controlled today using pesticides, actually became pests because pesticide use reduced or eliminated natural predators. A method of increasing grower adoption of biocontrol involves is letting growers learn by doing, for example showing them simple field experiments, having observations of live predation of pests, or collections of parasitised pests. In the Philippines, early season sprays against leaf folder caterpillars were common practice, but growers were asked to follow a 'rule of thumb' of not spraying against leaf folders for the first 30 days after transplanting; participation in this resulted in a reduction of insecticide use by 1/3 and a change in grower perception of insecticide use.

Companion Planting

Companion planting in gardening and agriculture is the planting of different crops in proximity for pest control, pollination, providing habitat for beneficial creatures, maximizing use of space, and to otherwise increase crop productivity. Companion planting is a form of polyculture.

Companion planting is used by farmers and gardeners in both industrialized and developing countries for many reasons. Many of the modern principles of companion planting were present many centuries ago in cottage gardens in England and forest gardens in Asia, and thousands of years ago in Mesoamerica.

History

In China, mosquito ferns (*Azolla* spp.) have been used for at least a thousand years as companion plants for rice crops. They host a cyanobacterium that fixes nitrogen from the atmosphere, and they block light from plants that would compete with the rice.

Companion planting was practiced in various forms by the indigenous peoples of the Americas prior to the arrival of Europeans. These peoples domesticated squash 8,000 to 10,000 years ago, then maize, then common beans, forming the Three Sisters agricultural technique. The cornstalk served as a trellis for the beans to climb, and the beans fixed nitrogen, benefitting the maize.

Companion planting was widely promoted in the 1970s as part of the organic gardening movement. It was encouraged for pragmatic reasons, such as natural trellising, but mainly with the idea that different species of plant may thrive more when close together. It is also a technique frequently used in permaculture, together with mulching, polyculture, and changing of crops.

Examples of Companion Plants

Nasturtium (*Tropaeolum majus*) is a food plant of some caterpillars which feed primarily on members of the cabbage family (brassicas), and some gardeners claim that planting them around brassicas protects the food crops from damage, as eggs of the pests are preferentially laid on the nasturtium. This practice is called trap cropping.

The smell of the foliage of marigolds is claimed to deter aphids from feeding on neighbouring crops. Marigolds with simple flowers also attract nectar-feeding adult hoverflies, the larvae of which are predators of aphids.

Various legume crops benefit from being commingled with a grassy nurse crop. For example, common vetch or hairy vetch is planted together with rye or winter wheat to make a good cover crop or green manure (or both).

Versions

There are a number of systems and ideas using companion planting.

Square foot gardening attempts to protect plants from many normal gardening problems by packing them as closely together as possible, which is facilitated by using companion plants, which can be closer together than normal.

Another system using companion planting is the forest garden, where companion plants are intermingled to create an actual ecosystem, emulating the interaction of up to seven levels of plants in a forest or woodland.

Organic gardening may make use of companion planting, since many synthetic means of fertilizing, weed reduction and pest control are forbidden.

Host-finding Disruption

Recent studies on host-plant finding have shown that flying pests are far less successful if their host-plants are surrounded by any other plant or even "decoy-plants" made of green plastic, cardboard, or any other green material.

The host-plant finding process occurs in phases:

The first phase is stimulation by odours characteristic to the host-plant. This induces the insect to try to land on the plant it seeks. But insects avoid landing on brown (bare) soil. So if only the host-plant is present, the insects will quasi-systematically find it by simply landing on the only green thing around. This is called (from the point of view of the insect) "appropriate landing". When it does an "inappropriate landing", it flies off to any other nearby patch of green. It eventually leaves the area if there are too many 'inappropriate' landings.

The second phase of host-plant finding is for the insect to make short flights from leaf to leaf to assess the plant's overall suitability. The number of leaf-to-leaf flights varies according to the insect species and to the host-plant stimulus received from each leaf. The insect must accumulate sufficient stimuli from the host-plant to lay eggs; so it must make a certain number of consecutive 'appropriate' landings. Hence if it makes an 'inappropriate landing', the assessment of that plant is negative, and the insect must start the process anew.

Thus it was shown that clover used as a ground cover had the same disruptive effect on eight pest species from four different insect orders. An experiment showed that 36% of cabbage root flies laid eggs beside cabbages growing in bare soil (which resulted in no crop), compared to only 7% beside cabbages growing in clover (which allowed a good crop). Simple decoys made of green cardboard also disrupted appropriate landings just as well as did the live ground cover.

Companion Plant Categories

The use of companion planting can be of benefit to the grower in a number of different ways, including:

- Hedged investment – the growing of different crops in the same space increases the odds of some yield being given, even if one crop fails.
- Increased level interaction – when crops are grown on different levels in the same space, such as providing ground cover or one crop working as a trellis for another, the overall yield of a plot may be increased.
- Protective shelter is when one type of plant may serve as a wind break or provide shade for another.
- Pest suppression – some companion plants may help prevent pest insects or pathogenic fungi from damaging the crop, through chemical means.
- Predator recruitment and positive hosting – The use of companion plants that produce copious nectar or pollen in a vegetable garden (insectary plants) may help encourage higher populations of beneficial insects that control pests, as some beneficial predatory insects only consume pests in their larval form and are nectar or pollen feeders in their adult form.
- Trap cropping – some companion plants are claimed to attract pests away from others.
- Pattern disruption – in a monoculture pests spread easily from one crop plant to the next, whereas such easy progress may be disrupted by surrounding companion plants of a different type.

Crop Rotation

Crop rotation is the practice of growing a series of dissimilar or different types of crops in the same area in sequenced seasons.It is done so that the soil of farms is not used to only one type of nutrient. It helps in reducing soil erosion and increases soil fertility and crop yield.

Growing the same crop in the same place for many years in a row disproportionately depletes the soil of certain nutrients. With rotation, a crop that leaches the soil of one kind of nutrient is followed during the next growing season by a dissimilar crop that returns that nutrient to the soil or draws a different ratio of nutrients. In addition, crop rotation mitigates the buildup of pathogens and pests that often occurs when one species is continuously cropped, and can also improve soil structure and fertility by increasing biomass from varied root structures.

Crop rotation is used in both conventional and organic farming systems.

History

It has long been recognized that suitable rotations – such as planting spring crops for livestock in place of grains for human consumption – make it possible to restore or to maintain a productive soil. Middle Eastern farmers practiced crop rotation in 6000 BC without understanding the chemistry, alternately planting legumes and cereals. In the Bible chapter of Leviticus 25, God instructs the Israelites to observe a 'Sabbath of the Land'. Every seventh year they would not till, prune or even control insects. The Roman writer, Cato the Elder, recommended that farmers "save carefully goat, sheep, cattle, and all other dung". In Europe, since the times of Charlemagne, there was a transition from a two-field crop rotation to a three-field crop rotation. Under a two-field rotation, half the land was planted in a year, while the other half lay fallow. Then, in the next year, the two fields were reversed.

From the end of the Middle Ages until the 20th century, three-year rotation was practiced by farmers in Europe. Under three-field rotation, the land was divided into three parts. One section was planted in the autumn with rye or winter wheat, followed by spring oats or barley, or other crops such as peas, lentils, or beans and the third field was left fallow. The three fields were rotated in this manner so that every three years, a field would rest and be fallow. Under the two-field system, if one has a total of 600 acres (2.4 km^2) of fertile land, one would only plant 300 acres. Under the new three-field rotation system, one would plant (and therefore harvest) 400 acres. But, the additional crops had a more significant effect than mere productivity. Since the spring crops were mostly legumes, they increased the overall nutrition of the people of Northern Europe.

A four-field rotation was pioneered by farmers, namely in the region Waasland in the early 16th century and popularised by the British agriculturist Charles Townshend in the 18th century. The system (wheat, turnips, barley and clover), opened up a fodder crop and grazing crop allowing livestock to be bred year-round. The four-field crop rotation was a key development in the British Agricultural Revolution.

George Washington Carver studied crop rotation methods in the United States, teaching southern farmers to rotate soil-depleting crops like cotton with soil-enriching crops like peanuts and peas.

In the Green Revolution, the traditional practice of crop rotation gave way in some parts of the world to the practice of supplementing the chemical inputs to the soil through top dressing with fertilizers, e.g. adding ammonium nitrate or urea and restoring soil pH with lime in the search for increased yields, preparing soil for specialist crops, and seeking to reduce waste and inefficiency by simplifying planting and harvesting.

Crop Choice

A preliminary assessment of crop interrelationships can be found in how each crop: (1) contributes to soil organic matter (SOM) content, (2) provides for pest management, (3) manages deficient or excess nutrients, and (4) how it contributes to or controls for soil erosion.

Crop choice is often a related to the goal the farmer is looking to achieve with the rotation, which could be weed management, increasing available nitrogen in the soil, controlling for erosion, or increasing soil structure and biomass, to name a few. When discussing crop rotations, crops are classified in different ways depending on what quality is being assessed: by family, by nutrient needs/benefits, and/or by profitability (i.e. cash crop versus cover crop). For example, giving adequate attention to plant family is essential to mitigating pests and pathogens. However, many farmers have success managing rotations by planning sequencing and cover crops around desirable cash crops. The following is a simplified classification based on crop quality and purpose.

Row Crops

Many crops which are critical for the market, like vegetables, are row crops (that is, grown in tight rows). While often the most profitable for farmers, these crops are more taxing on the soil Row crops typically have low biomass and shallow roots: this means the plant contributes low residue to the surrounding soil and has limited effects on structure. With much of the soil around the plant is exposed to disruption by rainfall and traffic, fields with row crops experience faster break down of organic matter by microbes, leaving fewer nutrients for future plants.

In short, while these crops may be profitable for the farm, they are nutrient depleting. Crop rotation practices exist to strike a balance between short-term profitability and long-term productivity.

Legumes

A great advantage of crop rotation comes from the interrelationship of nitrogen fixing-crops with nitrogen demanding crops. Legumes, like alfalfa and clover, collect available nitrogen from the soil in nodules on their root structure. When the plant is harvested, the biomass of uncollected roots breaks down, making the stored nitrogen available to future crops. Legumes are also a valued green manure: a crop that collects nutrients and fixes them at soil depths accessible to future crops.

In addition, legumes have heavy tap roots that burrow deep into the ground, lifting soil for better tilth and absorption of water.

Grasses and Cereals

Cereal and grasses are frequent cover crops because of the many advantages they supply to soil quality and structure. The dense and far-reaching root systems give ample structure to surrounding soil and provide significant biomass for soil organic matter.

Grasses and cereals are key in weed management as they compete with undesired plants for soil space and nutrients.

Green Manure

Green manure is a crop that is mixed into the soil. Both nitrogen-fixing legumes and nutrient scavengers, like grasses, can be used as green manure. Green manure of legumes is an excellent source of nitrogen, especially for organic systems, however, legume biomass doesn't contribute to lasting soil organic matter like grasses do.

Planning a Rotation

There are numerous factors that must be taken into consideration when planning a crop rotation. Planning an effective rotation requires weighing fixed and fluctuating production circumstances, including, but not limited to: market, farm size, labor supply, climate, soil type, growing practices, etc. Moreover, a crop rotation must consider in what condition one crop will leave the soil for the succeeding crop and how one crop can be seeded with another crop. For example, a nitrogen-fixing crop, like a legume, should always proceed a nitrogen depleting one; similarly, a low residue crop (i.e. a crop with low biomass) should be offset with a high biomass cover crop, like a mixture of grasses and legumes.

There is no limit to the number of crops that can be used in a rotation, or the amount of time a rotation takes to complete. Decisions about rotations are made years prior, seasons prior, or even at the very last minute when an opportunity to increase profits or soil quality presents itself. In short, there is no singular formula for rotation, but many considerations to take into account.

Implementation

Crop rotation systems may be enriched by the influences of other practices such as the addition of livestock and manure, intercropping or multiple cropping, and organic management low in pesticides and synthetic fertilizers.

Incorporation of Livestock

Introducing livestock makes the most efficient use of critical sod and cover crops; livestock (through manure) are able to distribute the nutrients in these crops throughout the soil rather than removing nutrients from the farm through the sale of hay. In systems where use of farm livestock would violate reservations growers or consumers may have about animal exploitation, efforts are made to surrogate this input through livestock in the soil, namely worms and microorganisms.

In Sub-Saharan Africa, as animal husbandry becomes less of a nomadic practice many herders have begun integrating crop production into their practice. This is known as mixed farming, or the practice of crop cultivation with the incorporation of raising cattle, sheep and/or goats by the same economic entity, is increasingly common. This interaction between the animal, the land and the crops are being done on a small scale all across this region. Crop residues provide animal feed, while the animals provide manure for replenishing crop nutrients and draft power. Both processes are extremely important in this region of the world as it is expensive and logistically unfeasible to transport in synthetic fertilizers and large-scale machinery. As an additional benefit, the cattle, sheep and/or goat provide milk and can act as a cash crop in the times of economic hardship.

Organic Farming

Crop rotation is a required practice in order for a farm to receive organic certification in the United States. The "Crop Rotation Practice Standard" for the National Organic Program under the U.S. Code of Federal Regulations, section §205.205, states that:

Farmers are required to implement a crop rotation that maintains or builds soil organic matter, works to control pests, manages and conserves nutrients, and protects against erosion. Producers of perennial crops that aren't rotated may utilize other practices, such as cover crops, to maintain soil health.

In addition to lowering the need for inputs by controlling for pests and weeds and increasing available nutrients, crop rotation helps organic growers increase the amount of biodiversity on their farms. Biodiversity is also a requirement of organic certification, however, there are no rules in place to regulate or reinforce this standard. Increasing the biodiversity of crops has beneficial effects on the surrounding ecosystem and can host a greater diversity of fauna, insects, and beneficial microorganism in the soil.< Some studies point to increased nutrient availability from crop rotation under organic systems compared to conventional practices as organic practices are less likely to inhibit of beneficial microbes in soil organic matter.

While multiple cropping and intercropping benefit from many of the same principals as crop rotation, they do not satisfy the requirement under the NOP.

Intercropping

Multiple cropping systems, such as intercropping or companion planting, offer more diversity and complexity within the same season or rotation, for example the three sisters. An example of companion planting is the inter-planting of corn with pole beans and vining squash or pumpkins. In this system, the beans provide nitrogen; the corn provides support for the beans and a "screen" against squash vine borer; the vining squash provides a weed suppressive canopy and discourages corn-hungry raccoons.

Double-cropping is common where two crops, typically of different species, are grown sequentially in the same growing season, or where one crop (e.g. vegetable) is grown continuously with a cover crop (e.g. wheat). This is advantageous for small farms, who often cannot afford to leave cover crops to replenish the soil for extended periods of time, as larger farms can. When multiple cropping is implemented on small farms, these systems can maximize benefits of crop rotation on available land resources.

Benefits

Agronomists describe the benefits to yield in rotated crops as "The Rotation Effect". There are many found benefits of rotation systems: however, there is no specific scientific basis for the sometimes 10-25% yield increase in a crop grown in rotation versus monoculture. The factors related to the increase are simply described as alleviation of the negative factors of monoculture cropping systems. Explanations due to improved nutrition; pest, pathogen, and weed stress reduction; and improved soil structure have been found in some cases to be correlated, but causation has not been determined for the majority of cropping systems.

Other benefits of rotation cropping systems include production cost advantages. Overall financial risks are more widely distributed over more diverse production of crops and/or livestock. Less reliance is placed on purchased inputs and over time crops can maintain production goals with fewer inputs. This in tandem with greater short and long term yields makes rotation a powerful tool for improving agricultural systems.

Soil Organic Matter

The use of different species in rotation allows for increased soil organic matter (SOM), greater soil structure, and improvement of the chemical and biological soil environment for crops. With more SOM, water infiltration and retention improves, providing increased drought tolerance and decreased erosion.

Soil organic matter is a mix of decaying material from biomass with active microorganisms. Crop rotation, by nature, increases exposure to biomass from sod, green manure, and a various other plant debris. The reduced need for intensive tillage under crop rotation allows biomass aggregation to lead to greater nutrient retention and utilization, decreasing the need for added nutrients. With tillage, disruption and oxidation of soil creates a less conducive environment for diversity and proliferation of microorganisms in the soil. These microorganisms are what make nutrients available to plants. So, where "active" soil organic matter is a key to productive soil, soil with low microbial activity provides significantly fewer nutrients to plants; this is true even though the quantity of biomass left in the soil may be the same.

Soil microorganisms also decrease pathogen and pest activity through competition. In addition, plants produce root exudates and other chemicals which manipulate their soil environment as well as their weed environment. Thus rotation allows increased yields from nutrient availability but also alleviation of allelopathy and competitive weed environments.

Carbon Sequestration

Studies have shown that crop rotations greatly increase soil organic carbon (SOC) content, the main constituent of soil organic matter. Carbon, along with hydrogen and oxygen, is a macronutrient for plants. Highly diverse rotations spanning long periods of time have shown to be even more effective in increasing SOC, while soil disturbances (e.g. from tillage) are responsible for exponential decline in SOC levels. In Brazil, conservation to no-till methods combined with intensive crop rotations has been shown an SOC sequestration rate of 0.41 tonnes per hectare per year.

In addition to enhancing crop productivity, sequestration of atmospheric carbon has great implications in reducing rates of climate change by removing carbon dioxide from the air.

Nitrogen Fixing

Rotating crops adds nutrients to the soil. Legumes, plants of the family Fabaceae, for instance, have nodules on their roots which contain nitrogen-fixing bacteria called rhizobia. It therefore makes good sense agriculturally to alternate them with cereals (family Poaceae) and other plants that require nitrates.

Pathogen and Pest Control

Crop rotation is also used to control pests and diseases that can become established in the soil over time. The changing of crops in a sequence decreases the population level of pests by (1) interrupting pest life cycles and (2) interrupting pest habitat. Plants within the same taxonomic family tend to have similar pests and pathogens. By regularly changing crops and keeping the soil occupied by cover crops instead of lying fallow, pest cycles can be broken or limited, especially cycles that benefit from overwintering in residue. For example, root-knot nematode is a serious problem for some plants in warm climates and sandy soils, where it slowly builds up to high levels in the soil, and can severely damage plant productivity by cutting off circulation from the plant roots. Growing a crop that is not a host for root-knot nematode for one season greatly reduces the level of the nematode in the soil, thus making it possible to grow a susceptible crop the following season without needing soil fumigation.

This principle is of particular use in organic farming, where pest control must be achieved without synthetic pesticides.

Weed Management

Integrating certain crops, especially cover crops, into crop rotations is of particular value to weed management. These crops crowd out weed through competition. In addition, the sod and compost from cover crops and green manure slows the growth of what weeds are still able to make it through the soil, giving the crops further competitive advantage. By removing slowing the growth and proliferation of weeds while cover crops are cultivated, farmers greatly reduce the presence of weeds for future crops, including shallow rooted and row crops, which are less resistant to weeds. Cover crops are, therefore, considered conservation crops because they protect otherwise fallow land from becoming overrun with weeds.

This system has advantages over other common practices for weeds management, such as tillage. Tillage is meant to inhibit growth of weeds by overturning the soil; however, this has a countering effect of exposing weed seeds that may have gotten buried and burying valuable crop seeds. Under crop rotation, the number of viable seeds in the soil is reduced through the reduction of the weed population.

Preventing Soil Erosion

Crop rotation can significantly reduce the amount of soil lost from erosion by water. In areas that are highly susceptible to erosion, farm management practices such as zero and reduced tillage can be supplemented with specific crop rotation methods to reduce raindrop impact, sediment detachment, sediment transport, surface runoff, and soil loss.

Protection against soil loss is maximized with rotation methods that leave the greatest mass of crop stubble (plant residue left after harvest) on top of the soil. Stubble cover in contact with the soil minimizes erosion from water by reducing overland flow velocity, stream power, and thus the ability of the water to detach and transport sediment. Soil Erosion and Cill prevent the disruption and detachment of soil aggregates that cause macropores to block, infiltration to decline, and runoff to increase. This significantly improves the resilience of soils when subjected to periods of erosion and stress.

The effect of crop rotation on erosion control varies by climate. In regions under relatively consistent climate conditions, where annual rainfall and temperature levels are assumed, rigid crop rotations can produce sufficient plant growth and soil cover. In regions where climate conditions are less predictable, and unexpected periods of rain and drought may occur, a more flexible approach for soil cover by crop rotation is necessary. An opportunity cropping system promotes adequate soil cover under these erratic climate conditions. In an opportunity cropping system, crops are grown when soil water is adequate and there is a reliable sowing window. This form of cropping system is likely to produce better soil cover than a rigid crop rotation because crops are only sown under optimal conditions, whereas rigid systems are not necessarily sown in the best conditions available.

Crop rotations also affect the timing and length of when a field is subject to fallow. This is very important because depending on a particular region's climate, a field could be the most vulnerable to erosion when it is under fallow. Efficient fallow management is an essential part of reducing erosion in a crop rotation system. Zero tillage is a fundamental management practice that promotes crop stubble retention under longer unplanned fallows when crops cannot be planted. Such management practices that succeed in retaining suitable soil cover in areas under fallow will ultimately reduce soil loss.

Biodiversity

Increasing the biodiversity of crops has beneficial effects on the surrounding ecosystem and can host a greater diversity of fauna, insects, and beneficial microorganisms in the soil. Some studies point to increased nutrient availability from crop rotation under organic systems compared to conventional practices as organic practices are less likely to inhibit of beneficial microbes in soil organic matter, such as arbuscular mycorrhizae, which increase nutrient uptake in plants. Increasing biodiversity also increases the resilience of agro-ecological systems.

Farm Productivity

Crop rotation contributes to increased yields through improved soil nutrition. By requiring planting and harvesting of different crops at different times, more land can be farmed with the same amount of machinery and labour.

Risk Management

Different crops can reduce the risks of adverse weather for the individual farmer.

Challenges

While crop rotation requires a great deal of planning, crop choice must respond to a number of fixed conditions (soil type, topography, climate, and irrigation) in addition to conditions that may change dramatically from year to the next (weather, market, labor supply). In this way, it is unwise to plan to crops years in advance. Improper implementation of a crop rotation plan may lead to imbalances in the soil nutrient composition or a buildup of pathogens affecting a critical crop. The consequences of faulty rotation may take years to become apparent even to experienced soil scientists and can take just as long to correct.

Many challenges exist within the practices associated with crop rotation. For example, green manure from legumes can lead to an invasion of snails or slugs and the decay from green manure can occasionally suppress the growth of other crops.

Integrated Pest Management

Integrated pest management (IPM), also known as integrated pest control (IPC) is a broad-based approach that integrates practices for economic control of pests. IPM aims to suppress pest populations below the economic injury level (EIL). The UN's Food and Agriculture Organisation defines IPM as "the careful consideration of all available pest control techniques and subsequent integration of appropriate measures that discourage the development of pest populations and keep pesticides and other interventions to levels that are economically justified and reduce or minimize risks to human health and the environment. IPM emphasizes the growth of a healthy crop with the least possible disruption to agro-ecosystems and encourages natural pest control mechanisms." Entomologists and ecologists have urged the adoption of IPM pest control since the 1970s. IPM allows for safer pest control. This includes managing insects, plant pathogens and weeds.

Globalization and increased mobility often allow increasing numbers of invasive species to cross national borders. IPM poses the least risks while maximizing benefits and reducing costs.

For their leadership in developing and spreading IPM worldwide, Perry Adkisson and Ray F. Smith received the 1997 World Food Prize.

History

Shortly after World War II, when synthetic insecticides became widely available, entomologists in California developed the concept of "supervised insect control". Around the same time, entomologists in the US Cotton Belt were advocating a similar approach. Under this scheme, insect control was "supervised" by qualified entomologists and insecticide applications were based on conclusions reached from periodic monitoring of pest and natural-enemy populations. This was viewed as an alternative to calendar-based programs. Supervised control was based on knowledge of the ecology and analysis of projected trends in pest and natural-enemy populations.

Supervised control formed much of the conceptual basis for the "integrated control" that University of California entomologists articulated in the 1950s. Integrated control sought to identify the best mix of chemical and biological controls for a given insect pest. Chemical insecticides were to be used in the manner least disruptive to biological control. The term "integrated" was thus synonymous with "compatible." Chemical controls were to be applied only after regular monitoring indicated that a pest population had reached a level (the economic threshold) that required treatment to prevent the population from reaching a level (the economic injury level) at which economic losses would exceed the cost of the control measures.

IPM extended the concept of integrated control to all classes of pests and was expanded to include all tactics. Controls such as pesticides were to be applied as in integrated control, but these now had to be compatible with tactics for all classes of pests. Other tactics, such as host-plant resistance and cultural manipulations, became part of the IPM framework. IPM combined entomologists, plant pathologists, nematologists and weed scientists.

In the United States, IPM was formulated into national policy in February 1972 when President Richard Nixon directed federal agencies to take steps to advance the application of IPM in all relevant sectors. In 1979, President Jimmy Carter established an interagency IPM Coordinating Committee to ensure development and implementation of IPM practices.

Applications

IPM is used in agriculture, horticulture, human habitations, preventive conservation and general pest control, including structural pest management, turf pest management and ornamental pest management.

Principles

An American IPM system is designed around six basic components:

Acceptable pest levels—The emphasis is on *control*, not *eradication*. IPM holds that wiping out an entire pest population is often impossible, and the attempt can be expensive and unsafe. IPM programmes first work to establish acceptable pest levels, called action thresholds, and apply controls if those thresholds are crossed. These thresholds are pest and site specific, meaning that it may be acceptable at one site to have a weed such as white clover, but not at another site. Allowing a pest population to survive at a reasonable threshold reduces selection pressure. This lowers the rate at which a pest develops resistance to a control, because if almost all pests are killed then those that have resistance will provide the genetic basis of the future population. Retaining a significant number unresistant specimens dilutes the prevalence of any resistant genes that appear. Similarly, the repeated use of a single class of controls will create pest populations that are more resistant to that class, whereas alternating among classes helps prevent this.

Preventive cultural practices—Selecting varieties best for local growing conditions and maintaining healthy crops is the first line of defense. Plant quarantine and 'cultural techniques' such as crop sanitation are next, e.g., removal of diseased plants, and cleaning pruning shears to prevent spread of infections. Beneficial fungi and bacteria are added to the potting media of horticultural crops vulnerable to root diseases, greatly reducing the need for fungicides.

Monitoring—Regular observation is critically important. Observation is broken into inspection and identification. Visual inspection, insect and spore traps, and other methods are used to monitor pest levels. Record-keeping is essential, as is a thorough knowledge target pest behavior and reproductive cycles. Since insects are cold-blooded, their physical development is dependent on area temperatures. Many insects have had their development cycles modeled in terms of degree-days. The degree days of an environment determines the optimal time for a specific insect outbreak. Plant pathogens follow similar patterns of response to weather and season.

Mechanical controls—Should a pest reach an unacceptable level, mechanical methods are the first options. They include simple hand-picking, barriers, traps, vacuuming and tillage to disrupt breeding.

Biological controls—Natural biological processes and materials can provide control, with acceptable environmental impact, and often at lower cost. The main approach is to promote beneficial insects that eat or parasitize target pests. Biological insecticides, derived from naturally occurring microorganisms (*e.g.—Bt*, entomopathogenic fungi and entomopathogenic nematodes), also fall in this category. Further 'biology-based' or 'ecological' techniques are under evaluation.

Responsible use—Synthetic pesticides are used as required and often only at specific times in a pest's life cycle. Many newer pesticides are derived from plants or naturally occurring substances (*e.g.*—nicotine, pyrethrum and insect juvenile hormone analogues), but the toxophore or active component may be altered to provide increased biological activity or stability. Applications of pesticides must reach their intended targets. Matching the application technique to the crop, the pest, and the pesticide is critical. The use of low-volume spray equipment reduces overall pesticide use and labor cost.

An IPM regime can be simple or sophisticated. Historically, the main focus of IPM programmes was on agricultural insect pests. Although originally developed for agricultural pest management, IPM programmes are now developed to encompass diseases, weeds and other pests that interfere with management objectives for sites such as residential and commercial structures, lawn and turf areas, and home and community gardens.

Process

IPM is the selection and use of pest control actions that will ensure favourable economic, ecological and social consequences and is applicable to most agricultural, public health and amenity pest management situations. The IPM process starts with monitoring, which includes inspection and identification, followed by the establishment of economic injury levels. The economic injury levels set the economic threshold level. That is the point when pest damage (and the benefits of treating the pest) exceed the cost of treatment. This can also be an action threshold level for determining an unacceptable level that is not tied to economic injury. Action thresholds are more common in structural pest management and economic injury levels in classic agricultural pest management. An example of an action threshold is one fly in a hospital operating room is not acceptable, but one fly in a pet kennel would be acceptable. Once a threshold has been crossed by the pest population action steps need to be taken to reduce and control the pest. Integrated pest management employ a variety of actions including cultural controls, including physical barriers, biological controls, including adding and conserving natural predators and enemies to the pest, and finally chemical controls or pesticides. Reliance on knowledge, experience, observation and integration of multiple techniques makes IPM appropriate for organic farming (excluding synthetic pesticides). These may or may not include materials listed on the Organic Materials Review Institute (OMRI) Although the pesticides and particularly insecticides used in organic farming and organic gardening are generally safer than synthetic pesticides, they are not always more safe or environmentally friendly than synthetic pesticides and can cause harm. For conventional farms IPM can reduce human and environmental exposure to hazardous chemicals, and potentially lower overall costs.

Risk assessment usually includes four issues: 1) characterization of biological control agents, 2) health risks, 3) environmental risks and 4) efficacy.

Mistaken identification of a pest may result in ineffective actions. E.g., plant damage due to over-watering could be mistaken for fungal infection, since many fungal and viral infections arise under moist conditions.

Monitoring begins immediately, before the pest's activity becomes significant. Monitoring of agricultural pests includes tracking soil/planting media fertility and water quality. Overall plant health and resistance to pests is greatly influenced by pH, alkalinity, of dissolved mineral and Oxygen Reduction Potential. Many diseases are waterborne, spread directly by irrigation water and indirectly by splashing.

Once the pest is known, knowledge of its lifecycle provides the optimal intervention points. For example, weeds reproducing from last year's seed can be prevented with mulches and pre-emergent herbicide.

Pest-tolerant crops such as soybeans may not warrant interventions unless the pests are numerous or rapidly increasing. Intervention is warranted if the expected cost of damage by the pest is more than the cost of control. Health hazards may require intervention that is not warranted by economic considerations.

Specific sites may also have varying requirements. E.g., white clover may be acceptable on the sides of a tee box on a golf course, but unacceptable in the fairway where it could confuse the field of play.

Possible interventions include mechanical/physical, cultural, biological and chemical.

Mechanical/physical controls include picking pests off plants, or using netting or other material to exclude pests such as birds from grapes or rodents from structures. Cultural controls include keeping an area free of conducive conditions by removing waste or diseased plants, flooding, sanding, and the use of disease-resistant crop varieties. Biological controls are numerous. They include: conservation of natural predators or augmentation of natural predators, Sterile insect technique (SIT).

Augmentation, inoculative release and inundative release are different methods of biological control that affect the target pest in different ways. Augmentative control includes the periodic introduction of predators. With inundative release, predators are collected, mass-reared and periodically released in large numbers into the pest area. This is used for an immediate reduction in host populations, generally for annual crops, but is not suitable for long run use. With inoculative release a limited number of beneficial organisms are introduced at the start of the growing season. This strategy offers long term control as the organism's progeny affect pest populations throughout the season and is common in orchards. With seasonal inoculative release the beneficials are collected, mass-reared and released seasonally to maintain the beneficial population. This is commonly used in greenhouses. In America and other western countries, inundative releases are predominant, while Asia and the eastern Europe more commonly use inoculation and occasional introductions.

The Sterile insect technique (SIT) is an Area-Wide IPM program that introduces sterile male pests into the pest population to trick females into (unsuccessful) breeding encounters, providing a form of birth control and reducing reproduction rates. The biological controls mentioned above only appropriate in extreme cases, because in the introduction of new species, or supplementation of naturally occurring species can have detrimental ecosystem effects. Biological controls can be used to stop invasive species or pests, but they can become an introduction path for new pests.

Chemical controls include horticultural oils or the application of insecticides and herbicides. A Green Pest Management IPM program uses pesticides derived from plants, such as botanicals, or other naturally occurring materials.

Pesticides can be classified by their modes of action. Rotating among materials with different modes of action minimizes pest resistance.

Evaluation is the process of assessing whether the intervention was effective, whether it produced unacceptable side effects, whether to continue, revise or abandon the program.

Southeast Asia

The Green Revolution of the 1960s and '70s introduced sturdier plants that could support the heavier grain loads resulting from intensive fertilizer use. Pesticide imports by 11 Southeast Asian countries grew nearly sevenfold in value between 1990 and 2010, according to FAO statistics, with disastrous results. Rice farmers become accustomed to spraying soon after planting, triggered by signs of the leaf folder moth, which appears early in the growing season. It causes only superficial damage and doesn't reduce yields. In 1986, Indonesia banned 57 pesticides and completely stopped subsidizing their use. Progress was reversed in the 2000s, when growing production capacity, particularly in China, reduced prices. Rice production in Asia more than doubled. But it left farmers believing more is better—whether it's seed, fertilizer, or pesticides.

The brown planthopper (Nilaparvata lugens), the farmers' main target, has become increasingly resistant. Since 2008, outbreaks have devastated rice harvests throughout Asia, but not in the Mekong Delta. Reduced spraying allowed natural predators to neutralize planthoppers in Vietnam. In 2010 and 2011, massive planthopper outbreaks hit 400,000 hectares of Thai rice fields, causing losses of about $64 million. The Thai government is now pushing the "no spray in the first 40 days" approach.

By contrast early spraying kills frogs, spiders, wasps and dragonflies that prey on the later-arriving and dangerous planthopper and produced resistant strains. Planthoppers now require pesticide doses 500 times greater than originally. Overuse indiscriminately kills beneficial insects and decimates bird and amphibian populations. Pesticides are suspected of harming human health and became a common means for rural Asians to commit suicide.

In 2001, scientists challenged 950 Vietnamese farmers to try IPM. In one plot, each farmer grew rice using their usual amounts of seed and fertilizer, applying pesticide as they chose. In a nearby plot, less seed and fertilizer were used and no pesticides were applied for 40 days after planting. Yields from the experimental plots was as good or better and costs were lower, generating 8% to 10% more net income. The experiment led to the "three reductions, three gains" campaign, claiming that cutting the use of seed, fertilizer and pesticide would boost yield, quality and income. Posters, leaflets, TV commercials and a 2004 radio soap opera that featured a rice farmer who gradually accepted the changes. It didn't hurt that a 2006 planthopper outbreak hit farmers using insecticides harder than those who didn't. Mekong Delta farmers cut insecticide spraying from five times per crop cycle to zero to one.

The Plant Protection Center and the International Rice Research Institute (IRRI) have been encouraging farmers to grow flowers, okra and beans on rice paddy banks, instead of stripping vegetation, as was typical. The plants attract bees and a tiny wasp that eats planthopper eggs, while the vegetables diversify farm incomes.

Agriculture companies offer bundles of pesticides with seeds and fertilizer, with incentives for volume purchases. A proposed law in Vietnam requires licensing pesticide dealers and government approval of advertisements to prevent exaggerated claims. Insecticides that target other pests, such as Scirpophaga incertulas (stem borer), the larvae of moth species that feed on rice plants allegedly yield gains of 21% with proper use.

References

- Acorn, John (2007). Ladybugs of Alberta: Finding the Spots and Connecting the Dots. University of Alberta. p. 15. ISBN 978-0-88864-381-0
- Johnson, Sue Ellen; Charles L. Mohler, (2009). Crop Rotation on Organic Farms: A Planning Manual, NRAES 177. Ithica, NY: National Resource, Agriculture, and Engineering Services (NRAES). ISBN 978-1-933395-21-0
- Charles Perrings; Mark Herbert Williamson; Silvana Dalmazzone (1 January 2000). The Economics of Biological Invasions. Edward Elgar Publishing. ISBN 978-1-84064-378-7
- Consoli, Fernando L.; Parra, José Roberto Postali; Trichogramma, Roberto Antônio Zucchi (28 September 2010). Egg Parasitoids in Agroecosystems with Emphasis on. Springer. ISBN 978-1-4020-9110-0
- Rajeev K. Upadhyay; K.G. Mukerji; B. P. Chamola (30 November 2001). Biocontrol Potential and its Exploitation in Sustainable Agriculture: Volume 2: Insect Pests. Springer. pp. 261–. ISBN 978-0-306-46587-1
- J. C. van Lenteren (2003). Quality Control and Production of Biological Control Agents: Theory and Testing Procedures. CABI. ISBN 978-0-85199-836-7
- Baur, Fred. Insect Management for Food Storage and Processing. American Ass. of Cereal Chemists. pp. 162–165. ISBN 0-913250-38-4
- Capinera, John L. (October 2005). "Featured creatures:". University of Florida website - Department of Entomology and Nematology. University of Florida. Retrieved 7 June 2016
- "Moving on from the mongoose: the success of biological control in Hawai'i". Kia'i Moku. MISC. 18 April 2012. Retrieved 2 July 2016
- Dufour, Rex (July 2015). Tipsheet: Crop Rotation in Organic Farming Systems (Report). National Center for Appropriate Technology. Retrieved May 4, 2016
- Baldwin, Keith R. (June 2006). Crop Rotations on Organic Farms (PDF) (Report). Center for Environmental Farming Systems. Retrieved May 4, 2016
- Coleman, Pamela (November 2012). Guide for Organic Crop Producers (PDF) (Report). National Organic Program. Retrieved May 4, 2016
- Gegner, Lance; George Kuepper (August 2004). "Organic Crop Production Overview". National Center for Appropriate Technology. Retrieved May 4, 2016

Permissions

All chapters in this book are published with permission under the Creative Commons Attribution Share Alike License or equivalent. Every chapter published in this book has been scrutinized by our experts. Their significance has been extensively debated. The topics covered herein carry significant information for a comprehensive understanding. They may even be implemented as practical applications or may be referred to as a beginning point for further studies.

We would like to thank the editorial team for lending their expertise to make the book truly unique. They have played a crucial role in the development of this book. Without their invaluable contributions this book wouldn't have been possible. They have made vital efforts to compile up to date information on the varied aspects of this subject to make this book a valuable addition to the collection of many professionals and students.

This book was conceptualized with the vision of imparting up-to-date and integrated information in this field. To ensure the same, a matchless editorial board was set up. Every individual on the board went through rigorous rounds of assessment to prove their worth. After which they invested a large part of their time researching and compiling the most relevant data for our readers.

The editorial board has been involved in producing this book since its inception. They have spent rigorous hours researching and exploring the diverse topics which have resulted in the successful publishing of this book. They have passed on their knowledge of decades through this book. To expedite this challenging task, the publisher supported the team at every step. A small team of assistant editors was also appointed to further simplify the editing procedure and attain best results for the readers.

Apart from the editorial board, the designing team has also invested a significant amount of their time in understanding the subject and creating the most relevant covers. They scrutinized every image to scout for the most suitable representation of the subject and create an appropriate cover for the book.

The publishing team has been an ardent support to the editorial, designing and production team. Their endless efforts to recruit the best for this project, has resulted in the accomplishment of this book. They are a veteran in the field of academics and their pool of knowledge is as vast as their experience in printing. Their expertise and guidance has proved useful at every step. Their uncompromising quality standards have made this book an exceptional effort. Their encouragement from time to time has been an inspiration for everyone.

The publisher and the editorial board hope that this book will prove to be a valuable piece of knowledge for students, practitioners and scholars across the globe.

Index

www.ingramcontent.com/pod-product-compliance
Lightning Source LLC
LaVergne TN
LVHW080632210926
848485LV00003B/253